Kinship with Monkeys

THE HISTORICAL ECOLOGY SERIES

THE HISTORICAL ECOLOGY SERIES

William Balée, Carole Crumley, editors

This series explores the complex links between people and landscapes. Individuals and societies impact and change their environments, and they are in turn changed themselves by their surroundings. Drawing on scientific and humanistic scholarship, books in this series focus on environmental understanding and on temporal and spatial change. The series explores issues and develops concepts that help to preserve historical ecological experiences and hopes to derive lessons for today from other places and times.

Kinship with Monkeys

THE GUAJÁ FORAGERS OF EASTERN AMAZONIA

Loretta A. Cormier

With original illustrations
by James P. Cormier

COLUMBIA UNIVERSITY PRESS

NEW YORK

Columbia University Press
Publishers Since 1893
New York   Chichester, West Sussex

Copyright © 2003 Columbia University Press

Library of Congress Cataloging-in-Publication Data
Cormier, Loretta A.
    Kinship with monkeys : the Guajá foragers of eastern Amazonia /
Loretta A. Cormier ; with original illustrations by James P. Cormier.
        p. cm. — (The historical ecology series)
    Includes bibliographical references and index.
    ISBN 0–231–12524–0 (cloth : alk. paper) — ISBN 0–231–12525–9
(paper : alk. paper)
    1. Guajá Indians. 2. Guajá Indians–Ethnozoology. 3. Monkeys—
Brazil—Maranhäo (State). 4. Human-animal relationships—Brazil—
Maranhäo (State). I. Title II. Series.

F2520.1.G69C67 2003
639'.118'098121—dc21

                                                    2003043770

∞

Casebound editions of Columbia University Press books are printed
on permanent and durable acid-free paper.

Printed in the United States of America
c 10 9 8 7 6 5 4 3 2 1
p 10 9 8 7 6 5 4 3 2 1

CONTENTS

ACKNOWLEDGMENTS

I owe a tremendous debt of gratitude to many people who helped make this project possible. First and foremost, I would like to thank the Guajá people for welcoming me into their community and allowing me to share their lives. They were extraordinarily patient in teaching me about their culture, and I appreciate the kindness and friendship that they showed me, particularly Haikaramukã, Arakanei, Takamãči'a, Wiriči, and Iname. A number of people were very helpful to me in Brazil. I would like to thank the FUNAI (Fundação Nacional do Índio) officials and the CNPq (Conselho Nacional de Desenvolvimento Tecnológico e Científico) for their assistance with the project. The local FUNAI officials on the Caru reserve were extremely helpful in introducing me to the Guajá and providing information about their experiences with them. I would like to thank José A. Damasceno da Silva, Maria Dalva A. da Costa, Hilton Aguiar, Antônio Sales Oliveira, Raimundo Nonato Abreu Oliveira, and Olegário Ferreira Silva for the many ways they aided me. In Santa Inês, Renildo Matos dos Santos was very helpful, as were Fiorello Parise and Frederico de Miranda Oliveira in Belém. My Brazilian "orientador" at the Universidade Federal do Pará in Belém, João Farias Guerreiro, helped to facilitate aspects of the project locally. I also appreciate the assistance I received with plant identifications at the Museu Goeldi from Graça Lobo, Ricardo Secco, and Luís Carlos Batista Lobato. In addition, the staff at the Arquivo Público de Pará in Belém helped me obtain archival materials.

It is difficult to express how profoundly grateful I am to have had the opportunity to study under my anthropology professors. Those who particularly inspired me were William Balée, Victoria Bricker, Jane Christian, Margaret Clarke, John Hamer, Brian Hesse, Kathyrn Oths, John Verano, Elizabeth Watts, and Bruce Wheatley. I am particularly indebted to Bill Balée; I have benefited not only from his scholarship in ecological anthropology, but also from the relationships he established with the Guajá people, the FUNAI, and other professionals in Brazil, which were invaluable in opening doors for me to enter into the field. He, along with Anne Bradburn at the Tulane Herbarium, also provided much assistance with my plant collections after returning from the field. I would also like to extend a special thanks to Leslie Sponsel at the University of Hawaii for his advice in conducting fieldwork in ethnoprimatology. I would also like to extend thanks to the staff at the New Orleans Audubon Zoo, Larry Williams and Christian Abee at the University of South Alabama, and Dorothy Fragaszy at the University of Georgia, who allowed me to conduct preliminary research with their monkeys.

Financial support from several institutions was important in making this research possible. Primary funding was provided by a Fulbright Scholarship to Brazil. Additional funding or equipment came from the American Society of Primatologist's Small Conservation Grant, the Tinker Foundation, and a equipment donations from the Tulane Delta Regional Primate Center.

Finally, I am grateful for the support of my family, particularly my mother, Annette Garrison, who has helped me in many, many ways. Most of all, I would like to express my gratitude to my husband James. While in the field, he worked as a full-time partner in the research, contributing not only by taking time off from his own pursuits to accompany me, but helping me directly with the project in innumerable ways. He chopped firewood, made biscuits, scaled trees for plant samples, repeatedly rethatched our leaky hut roof, and even willingly helped hoist white-lipped peccary carcasses onto my hanging weight scales. I benefited enormously from his unique perspective and insights into the Guajá culture and from the relationships he cultivated with the men, which allowed me to gain information to which I might not have otherwise had easy access. The realization of this project is as much a product of his efforts as it is mine.

NOTES ON ORTHOGRAPHY

The following provides a brief, broad guide to Guajá pronounciation. Additional information on Guajá phonemics can be seen in Cunha's (1987) preliminary analysis.

VOWELS

i:	high, front vowel
ɨ:	high, central vowel
u:	high, back vowel
e:	mid, front vowel
ə:	mid, central vowel
o:	mid, back vowel
a:	low, back vowel

All vowels may also be nasalized, which is a phonemic difference. Nonnasalized final vowels may be pronouced as a dipthong with a mid, central vowel (i.e., iə, ɨə, uə, eə, oə, aə). Here, the initial sound of the dipthong is accented. This feature is in free variation, but some pronounciations are more commonly used. In the text, the most common pronounciations are used. For example:

/ ka'i / → [ka'i], [ka'íə]; the former pronounciation without the dipthong is more common

/ yu / → [yu], [yúə]; the latter pronounciation with the dipthong is more common

CONSONANTS

p: voiceless bilabial stop

When / p / is the initial consonant of a word, nasalization is in free variation. In positions other than the initial consonant of a word, vocalization is in free variation.

/ p / = [p], [b], [m] when the initial consonant of a word
/ p / = [p], [b] when in a position other than the initial consonant of a word

t: voiceless alveolar stop

A few informants pronouce this as an alveolar-palatal affricate when it is the initial sound of a word. When it occurs in a position other than the initial consonant of a word, vocalization is in free varation, although it is most commonly non-vocalized.

/ t / = [t], [č] when the initial consonant of a word
/ t / = [t], [d] when in position other that the initial consonant of a word

k: voiceless velar stop

Aspiration and vocalization are in free variation, regardless of the position of the consonant within the word. However, vocalization and aspiration are less commonly used.

/k/ = [k], [kh], [g]

':   glottal stop
h: glottal fricative

It is most commonly pronouced as a velar fricative but is sometimes pronounced as more of a glottal fricative when preceding central and back vowels.

č: voiceless palatal affricate

This sound may also be aspirated. Rarely, this is pronounced in free variation as a voiceless fricative when it is the initial consonant of a word.

/ č / = [č],[čʰ]
/ č / = [č], [š] when initial consonant of a word

m: voiced nasalized bilabial
n: voiced nasalized alveolar
r: voiced alveolar flap

This is sometimes pronounced with nasalization and is sometimes difficult to distinguish from / n /.

w: voiced bilabial glide
y: palatal-alveopalatal glide
kw: consonant cluster of voiceless velar stop and bilabial glide
ts: consonant cluster of voiceless alveolar stop and voiceless fricative

STRESS PATTERN

Stress is generally on the final syllable of the word. The most common exception to the rule involves words ending in a mid-central vowel / ə /. Often these words involve stress on the penultimate syllable, which is indicated in the text by an accent mark. For example:

Yawárə final mid-central vowel with stress on penultimate syllable
Harə final mid-central vowel with stress on final syllable

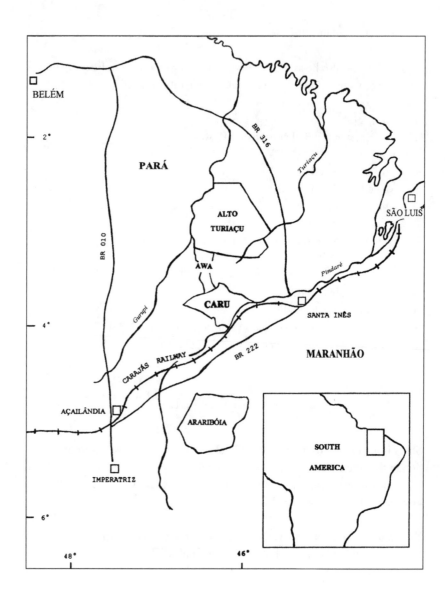

FIGURE 0.1 Indigenous areas of the Guajá

INTRODUCTION

ANTHROPOLOGY

One of my first anthropology professors, John Hamer, used to tell his students that you cannot make a person into an anthropologist; people are drawn to this field because on some level they already are anthropologists. I have thought about this idea over the years, and although I still am not sure exactly what that means for me, I do believe that it is true. I have heard anthropologists characterized as philosophers of human nature, pure scientists, humanitarians, thrill-seekers, or even as people attracted to other cultures because they already feel alienated from their own. For my part, I have to admit that my motives are somewhat selfish. Although the idea of experiencing another culture has always been rather intimidating, it is not as frightening as the idea of having lived my life without having made an attempt to see outside of my own narrow frame of reference.

These days in our field, critical self-reflection is considered by many to be as much a part of ethnography as is the study of the people themselves. Some would say this is for the better, some would say for the worse. Although I have certainly tried to present an objective, unbiased account of the Guajá people, I recognize that I am a product of my upbringing, my historical context, and my academic genealogy. I have to admit that I dread the thought of being deconstructed and dismissed in the future as a typical representative of the bias of whatever paradigm I cannot currently see beyond. Then again, I suppose I would be quite fortunate to be considered even worthy of the attention.

At the risk of sounding contrived, I will describe my research topic as coming to me in a kind of "eureka" moment. In my first semester as a graduate student at Tulane, Bill Balée showed a slide in an ecological anthropology course of a Guajá woman breast-feeding a monkey. I was both fascinated and horrified by the image, and could not stop thinking about how these people must view nature in order to have this kind of intimate relationship with monkeys. I had been torn early on in my education as to whether to be cultural anthropologist or a prima-tologist, and the Guajá seemed to present an ideal way to combine my interests. My interest had also been prefertilized by my first primatology professor, Bruce Wheatley, who had lectured to us about the critically neglected understanding of the effect of hunting on primate behavior and how this influenced his early in-terest in understanding the Balinese cultural conceptions of temple macaques.

Initially, I thought that the Guajá must perceive nature very differently than members of Western society do. We have been described as attempting to control nature—"man against nature," if you will. But the Guajá relationship to mon-keys suggested that perhaps they were nurturing nature itself by adopting this ma-ternal role toward monkeys. On the other hand, I thought that perhaps the Guajá were not all that different from us. Perhaps their pet monkeys were much like our pet dogs and cats, and I was simply having an ethnocentric reaction to the breast-feeding. However, as I read what research was available on the Guajá, the ques-tion became more complicated. Queiroz and Kipnis (1991) had done an archaeo-logical study of an old Guajá camp and found that monkeys were the primary game food of the Guajá. This apparent contradiction of the predatory/protector relationship of the Guajá and the monkeys became the central theme of my re-search: *How can monkeys be both eaten as food and nurtured as children?*

FIELDWORK AMONG THE GUAJÁ

The Guajá are a Tupi-Guarani speaking, traditional foraging people living in western Maranhão, Brazil. I spent approximately fifteen months among the Guajá between February 1996 and August 1997. In October 1996, I returned to the United States for several months before resuming field work in January 1997. My husband, James, who is not an anthropologist, accompanied me into the field, and we lived in a room in an infirmary built for the Guajá by the Brazilian Indian Agency, FUNAI (Fundação Nacional do Índio). It was no more than a hut consisting of a concrete floor, plank walls, and a thatched babassu palm roof that constantly leaked. We wound up pitching our tent inside of the hut, not so much for protection from the rain, but to guard against insects; I had been extremely naive about the insect load in the area. Although we meticulously cleaned the place, we had goliath tarantulas, scorpions, locusts, biting centipedes, and large lizards in our hut, in addition to a general infestation of cockroaches, gnats, and mosquitoes. The Guajá have falciparum malaria in their area, so mosquitoes were a serious concern. We also had ant invasions on several occasions by small black ants the Guajá call *tahɨ̃a* (*Eciton* sp.) that periodically swarmed our hut in masses of thousands. The tent was virtually waterproof and bugproof and seemed to work better than a hammock and mosquito nets. The tent also worked well against the *tahɨ̃a* ants, as they simply crawled over the tent during their night invasions.

Part of my daily routine included random spot checks, which involved visiting each of the ten Guajá huts in the vicinity at randomly selected times during the day and recording the residents' activities. Although I conducted these random spot-checks throughout the research period, much of the data I collected during the first few months was unusable because I did not understand the language well enough to really grasp their descriptions of those activities that were not obvious or familiar. But the process of the first random spot checks helped me acquire a feel for their daily routine and contributed to my learning the language. Household size ranged from six to fourteen individuals. Most of the houses were dirt-floor huts thatched by babassu palm leaves on all sides and on the roof. However, two of the huts had wattle and daub walls, similar to those used in the house construction of many local, non-Indian Brazilians.

During the first seven months, I was involved in two other structured research activities: focal animal samples of the monkeys and plant collecting. Obtaining focal animal samples involved observing pet monkeys in and around the Guajá households. In addition to gathering information about human-monkey interactions, this gave me an opportunity to observe many of the daily activities of the Guajá. Collecting the plant samples involved going into the forest, usually with two informants at a time, to identify and gather plants. Usually, the two would be a husband and wife team or two siblings. I marked the plants they identified and returned later to collect samples. When the foliage was located high in the canopy, sometimes the Guajá would be able to climb the trees and retrieve a sample. When it was out of reach of the Guajá, the method for retrieval that worked best involved a limb saw. James used a bow and arrow to shoot a length of string over a desired branch, which we would use to pull up a limb saw. Then pulled back and forth on ropes attached to either side of the limb saw to remove the sample.

Plant samples from the first part of the research period were shipped to the Museu Goeldi in Belém to be dried. On our return to the field, James constructed a plant drying oven that could be used over the hearth in our hut. We used a grill for the base of the oven, and the sides and top were metal sheets with temperature gauges on the tops and bottoms to monitor the heat level. We collected charcoal-like remains from our campfires as fuel so that we could keep a low, slow heat going to dry the plants.

During the second half of the research period, in addition to continuing with the random spot checks and the plant collections, I began weighing animals killed for food and conducting structured interviews with the Guajá. Informants were generally cooperative about bringing fish, birds, and game to be weighed in order to assess the relative importance of primates in the diet. By this part of the research, my language skills were sufficiently developed to begin conducting structured interviews in the Guajá language. I worked with pairs of informants and asked specific questions regarding kinship, religion, and other aspects of their culture. Although the structured research activities were important to the research, most of my information about the Guajá culture came from my interactions with them on a day-to-day basis, that is, through participant observation.

My husband, James, had many roles in the field. I benefited enormously from his being there and my having, essentially, a full-time partner to help. In practical terms, he took care of the bulk of our chores for daily survival, which gave me more time to focus on research. His relationships with the Guajá were more much more relaxed than mine. My interactions were oriented towards my research goals, while his could be more natural and casual. His friendships with the men were particularly beneficial in providing many insights that I might not have otherwise gained. Probably most important, his presence helped ease the cultural shock I felt while living with the Guajá. Sharing the experience with him made all aspects of the fieldwork much easier and much more enjoyable.

The Guajá are in transition from their traditional foraging lifestyle to a more settled existence. My ability to work among them the way that I did was facilitated by the rapid changes that have occurred in their way of life over approximately the last twenty years. I was not the first anthropologist to work among the Guajá, but they have less contact than most contemporary indigenous peoples due to their nomadic habits. Some Guajá have now been in contact with personnel from the Brazilian Indian Agency for much of their adult lives. However, some had their first contact only a few months before my arrival, and perhaps as many as one-third of them still remain isolated from non-Indians. Ironically, the very train that brought me within several hours of hiking and boating to the Caru reserve, where I worked, is largely responsible for the rapid changes that are now occurring. Construction began on the Carajás railroad in 1985; it runs from the coast of Saõ Luis to the interior where iron is mined. Although no mining is currently taking place in the Guajá area, the train runs through the middle of traditional Guajá territory, and in its wake it has brought illegal invasions, development, and disease. I know of no other way to describe the effect that this has had on the Guajá than to say that it has been devastating.

The most difficult part of the research was learning the Guajá language. No formal studies had been done of their language, and only a few of them knew any Portuguese. Those who spoke some Portuguese used a pidgin Portuguese-Guajá. Their pidgin typically involved substituting Portuguese terms that are easily glossed into Guajá, such as *peixe* ("*pira*") for fish and *guariba* ("*wari*") for howler.[1] So their pidgin is fundamentally Guajá.

I suspect that anthropologists often have the experience of finding themselves studied by the people they are studying. This was even more apparent for me with the Guajá who had not long been in contact. Shortly after we arrived, we met the group who had come into contact only a few months earlier. One young man named Takwarichi'a listened intently as James and I spoke to each other in English. I attempted to say a few words in Guajá to him, and he responded by saying politely to me, "Zzzzuh, zzzzuh, zzzzuh." This meant two things. First, that my early attempts at the Guajá language were obviously completely unintelligible to him. Secondly, he was hearing in our language the /z/ phoneme that was not present in his own. Apparently, he thought we communicated by making this meaningless buzzing sound, and so he did his best to replicate it and try to make contact.

On another occasion, still early in the field work, I was visiting in a Guajá hut making observations of their pet monkeys. A group of women were cooking and chatting to one another, and I attempted to join in the conversation. I do not remember the topic, but I managed to put together a sentence with some semblance of grammar. Immediately, one of the women shushed everyone around her and stared at me in amazement, encouraging me to speak again. I do not know exactly what she was thinking, but it seemed to me that it was probably the first time she considered the possibility that I could actually be an intelligent form of life.

When I first went to Brazil, I had basic conversational skills in Portuguese. I first began learning Guajá from several of the younger men who knew some Portuguese. Initially, I also learned some of the basic Guajá vocabulary from interactions with the children. They enjoyed playing a game with me that involved pointing to an that I was then supposed to try to name in Guajá while they tried to name it in Portuguese. The hardest people to learn to communicate with were the older women, none of whom knew any Portuguese. I eventually possessed sufficient language skills to understand the expression of indigenous concepts, but I never became fluent in Guajá.

A few months after arriving, I learned that the Guajá were being exposed to my society by more than my presence. One afternoon, a young girl named Iawowičika walked by my hut singing a reggae tune in nearly perfect English: "Bad boys, bad boys, whatcha gonna do? Whatcha gonna do when they come for you?" I rushed towards her with what I am sure was a crazed look on my face asking her to sing it again. Not surprisingly, my reaction scared her. I brought James over and convinced her to sing it to him, in part to prove to myself that I had not imagined it in some malarial haze. She explained, "televisão." We did not know it at the time, but the FUNAI workers at the post, about a mile away, had recently installed a satellite dish, and some of the Guajá were travelling there at night to watch. The "bad boys" song is from an American television show called COPS that had been dubbed, except for the theme song, into Portuguese. Although the words had no meaning to her, Iawowičika was able to mimic the sounds. Shortly thereafter, the children began playing a game called they called "policía." The child in the role of the police officer would chase other children and knock them down.

Needless to say, I was very disappointed that the Guajá were exposed to my own society in this way. I tried to explain that the police were the good guys, but this made no sense given what they saw on the show, which was the police chasing people, wrestling them to the ground, and apparently kidnapping them. Another game developed shortly afterwards called, *kamara* (non-Guajá Indians), which resulted from watching old Westerns. The children would fight each other riding sticks which were make-believe *tapi'ira* (tapirs, a distant relative of the horse). Next came the martial arts/karate-kicking game. The imitation of television violence was extremely disturbing. Currently in American society, there is much debate regarding the extent to which exposing children to television violence encourages subsequent violent behavior. The Guajá, like many hunter-gatherers, avoid physical conflict within the group. It is not completely absent, but it is rare. Interpersonal conflicts are dealt with primarily through joking and teasing each other. Typically, if a conflict cannot be resolved, it is avoided by simply moving away from each other into the forest. Although I did not study this behavior of the children in any controlled way, it appeared to be a clear indictment of the effects of television violence.

The material culture that we brought with us to the camp also influenced the Guajá in a way we had not anticipated. James and I were quite dismayed to find that an activity we thought was harmless was having a revitalization-movement effect among them. Our friends and relatives sent us care packages that included recent magazines; the Guajá children were fascinated with the pictures, and we frequently let them look through them. The adults seemed to take less of an interest, but did flip through them from time to time. The one exception to this was one very old Guajá woman, Merekechi'a, who repeatedly came over to look at a *Time* magazine picture of England's former prime minister, John Major. She said that he was *parahei* (beautiful) and wanted to know who he was. I told her that he was a grandfather, which was the closest translation I could think of that would make any sense in their acephalous society. I have wondered what made John Major stand out as beautiful among all the other faces she saw in the magazine.

While we were there, a photographer and writer visited the Guajá briefly for a feature insert in the São Luis newspaper. On one of our supply trips, we brought back a number of the inserts and gave them to the Guajá. Over a period of weeks, the adult men began to "read" these inserts. They stared at the pictures and chanted random Portuguese words and numbers. This was then generalized to all reading materials, and adults would come over, pick up a book or magazine, chant in Portuguese for several minutes, and then leave. I observed this for a while before I decided asked them to explain what they were doing. The reply was that they were doing the same thing we were doing: looking into the books to find answers, as they had seen us do many times, such as when looking up a word in our Portuguese dictionary. "Reading" became a form of divination practiced by some of the adult men. Some of the men had the ability to "read," while others were said to not be able to understand. One individual was deciding when adolescents were ready to make their first spiritual trancing journey to their sacred sky home by looking into the books and magazines.

The ritualization of Western media had actually begun before the arrival of the television and our books and magazines. Soon after we arrived, I observed adult men chanting Portuguese numbers or words while standing alone in the middle of the village or while walking down the trail. Initially I thought this was some type of ceremonial boasting. But eventually, James recognized that this peculiar form of speech for them sounded like the speech patterns of Brazilian radio and television broadcasters. We questioned the Guajá specifically on this, and it was indeed the case. Although somewhat reluctant to admit it, the men said that they were trying to send their voices to other parts of the forest, like a radio.

BECOMING HUMAN

The more I look back on my stay with the Guajá, the more I appreciate how extremely tolerant and patient they were with us. Overall, we were well received by the Guajá and got along well with them. I am sure that we seemed very peculiar, not only because our behavior was often so different from their own, but also because of our ineptitude at what must have seemed to them to be the very basic tasks of life. I frequently turned down the wrong path in the forest, I did not eat monkeys, I did not know how to make a hammock, and, perhaps most curious to them, I did not have children. We were questioned repeatedly about this: "Where are your children?" Typically, when I first told someone that I did not have children, he or she responded by expressing regret that my children had died. I tried to explain that I had never had children and that the people in my culture sometimes wait until they are older to have children or decide not have them at all. This did not seem to be a reasonable answer to them.

I was thirty-three when I began the fieldwork, and by Guajá standards, I should have been a grandmother by then. One day, several of them decided to question us further on this problem we seemed to be having. The Guajá, like many lowland South American groups, have a concept called "partible paternity." Here, fetuses are believed to be made from semen, and it is possible for a child to have more than one "biological" father. Among the Guajá, this is taken a step further, and they believe that multiple "fathers" are usually necessary to make a fetus. One of the Guajá in the group that came by asked us straight out whether we had been having sex. Considering our seeming lack of knowledge of so much of what is needed to get along in the world, I believe they wanted to make sure we understood that this was a necessary part of reproducing. I assured them that, yes, we did understand that part of it, and then another asked me, "Is it only with James?" And I said yes, in my culture, you are only supposed to have sex with one husband. I saw the knowing look pass between them. That, of course, was our problem. We did not understand that it takes more than one man to make a baby.

Another great cultural divide between us was our failure to abide by their cultural rule of sharing with anyone who asks. Although we did give small gifts in exchange for research tasks, we tried to be very careful about introducing too many Western material goods into their culture. It was difficult for us to deal with what at times seemed like incessant demands for Western goods. I do not think the Guajá ever really understood why we could not supply them with all their requests for machetes, knives, clothes, guns, and other items. Certainly, many thought we were rude and selfish to withhold things from them that they felt they needed. They tended to deal with this conflict with us in the typical Guajá way, by teasing us and making us the butt of their jokes. I do believe that we were a great source of entertainment for them. In one of the households, nearly every time I came by for my daily rounds, the husband and wife began laughing. And they really laughed, sometimes to the point of falling out of their hammocks. When I would ask what was so funny, they invariably replied that it was something that their monkey had just done. But I made no mistake; I knew that I was the real monkey here.

One thing that I was able to do that I believe helped to ameliorate their perception of me as selfish was to provide some health care. My undergraduate degree is in nursing, and I brought a few medical supplies with me. I quickly learned, though, that health care had to be given within their own cultural context. Most of their medicinal plants are categorized as such because they have a strong odor. They bathe in a solution made from these plants to ward off cannibal ghosts, which they believe cause illness. The most common health problems for the Guajá are respiratory viruses, and I would sometimes rub Vick's Vapor Rub under the children's noses and on their chests. Its strong smell categorized this as a potent medicine, (*puha*). I then ran into the problem of the Guajá wanting me to use Vick's Vapor Rub for a multitude of problems, such as headaches, diarrhea, and malarial fever. Initially, I told them that it did not work for that, but they did not believe me and seemed to interpret my answer as an unwillingness to help. So I began rubbing a little on whenever they asked and then also giving them a more effective medication.

Often, an important part of an anthropologist's acceptance into a culture involves receiving a kin term. In many cultures, a kinship designation is needed to guide the appropriate behavior towards an individual. For me, the day came when two small brothers came by, playing. They had learned the Portuguese word for mother (*mamãe*) from the FUNAI, and they started a game that involved dancing around me and singing over and over again, "*mamãe, mamãe, mamãe.*" And it stuck. Soon thereafter, everyone called me *mamãe*. It was an apt term for several reasons. The word for mother in their language, (*ama*), means not only mother, but, more generically, a woman accorded respect. Although the term was polite, using the Portuguese rather than the Guajá, suggested my ambiguous status as both an insider and an outsider.

However, on several occasions, our outsider status was reaffirmed to us. One such occasion was when I learned that the Guajá did not think that I had a soul. Their sacred sky home is interpreted as both the place where the soul goes when it dies and also the past. In their view, the souls of the Guajá on earth have counterparts in the sacred sky home in the past. Initially, I thought that the past in the sky represented the literal past. I was asking one of the grandfathers if something that I had been doing the day before existed in the sky home past. He shook his head sadly and said, "*Mamãe na'ači*" (you aren't there). It is not any past event, it is the sacred Guajá past. I was not Guajá, and therefore I was bound to the present time, with only a temporary, ephemeral existence.

Another, more tragic event that occurred during our stay also made clear our ambiguous status. Despite the Guajá norm of non-violence within the group, they are very capable of responding with deadly force to outside threats to the group. Non-Indians are referred to as "*karaí*." The Guajá reserve is now frequently invaded by squatters, loggers, and locals who come to the reserve to hunt and fish illegally. About seven months after we arrived in the field, three young Guajá men confronted a man and his wife, who had entered the reserve to fish. The Guajá told them to give up their fish and leave. When the man refused, one of the Guajá fired at him with a shotgun. As he pleaded for his life and tried to swim away, they chased after him; a second Guajá shot him with an arrow, and the third slit his throat with a machete.

Afterwards, the Guajá heard that more *karaí* were coming to the reserve to seek revenge. As we sat outside our hut, we watched the men run by with their bows and arrows, prepared to fight. Several ran past us without stopping. They did not consider us part of this. Another hesitated, stopped in front of our hut, started to go back home, paced a while, and then eventually ran on towards the others. He seemed to consider us *karaí* and was ambivalent about killing "one of us." Then another young man stopped and excitedly told us, "*Karaí u, karaí u*" (the *karaí* are coming). He showed us his arrows and encouraged us to come along with them. He considered us Guajá allies against the *karaí*.

Fortunately, the *karaí* were not coming. The Guajá had misunderstood a request from the FUNAI to come to the post and discuss the incident. Legally, the Guajá can kill any one who invades their reserve without consequences. The FUNAI workers merely wanted to try and understand what had happened and to encourage the Guajá to seek other alternatives. The local FUNAI were in an incredibly difficult situation. The man who was killed had been a friend of one of the FUNAI workers, and the others knew him well. The FUNAI workers make efforts to discourage the local people from coming onto the reserve, but many who live in the impoverished villages that have developed in the wake of the Carajás train come onto the reserve in acts of desperation. They are willing to risk their lives to feed their families. After the murder, the FUNAI workers received multiple death threats. The bravery of the FUNAI on the Caru is commendable, given their attempts to negotiate between the Guajá and the local Brazilians, as is their willingness to stay at all under such circumstances.

Later, on the day of the murder, each of the three men involved did come and talk with us. The first to come offered no excuses and simply came to reassure us that we were in no danger. This was not the first time he had killed a *karaí*. He had been involved in a similar incident several years earlier and hated the *karaí* with good reason. As a teenager, his family band had been massacred by a group of local Brazilians. Apparently, after his narrow escape, he just kept running, all the way to Minas Gerais, several states to the south. After he came into contact with local Brazilians there, he was eventually brought to the Caru reserve. I cannot pretend to understand the rage that he must feel, and I do not now how I would react under similar circumstances. Although he came to reassure me, his words had the opposite effect because before this I had not considered that we could be in any real danger from the Guajá.

The other two young men involved came together to talk with us. They were two of my main informants, whom I considered friends. They expressed regret at what had happened, but said they felt they had no choice. The *karaí* had been coming on the land and taking their game animals, their fish, and their trees. They felt that if they did not act to stop them, the Guajá would have nothing to eat. After that incident, the Guajá began explicitly referring to us as *"americanos,"* the term the FUNAI workers used for us. And we made a conscious effort to reinforce this because it helped to categorize us with a less culturally loaded term and avoided the problem of having *karaí* living in such close proximity to the Guajá. I would add though, that near the end of our stay, some began referring to us as *"awa-americano."* *"Awa"* is the term the Guajá use for themselves. It seemed that we had almost become human.

ETHNOPRIMATOLOGY

Monkeys are actually more human than anthropologists to the Guajá. In fact, so are the trees. In Guajá animistic beliefs, all forms of plant and animal life in their forest are endowed with souls and are woven into their kinship system. Among the forest beings, monkeys, and especially the howlers, are considered to be more closely related to the Guajá than any other forest being. Herein lies the resolution to the paradox of how monkeys can be both nurtured as children and hunted for food. According the Taylor (1996), in Amazonian thought, all death is homicide due to the belief in the involvement of some type of malignant human agency. This basic theme takes varied forms in different lowland South American cultures. Among the Guajá, forest life is anthropomorphized and all forest life is kin. Therefore, all consumption can be considered a form of cannibalism.

The role of monkeys in the Guajá culture is not merely alimentary. Infants whose mothers are killed for food are kept as pets and incorporated to an even greater extent into the kinship system at the household level. These pets are never eaten, are nurtured as children, and can even be considered to serve as surrogate children to Guajá women. Pet monkeys have a considerable presence in the Guajá community. In a village of approximately 108 individuals, 90 monkeys were kept as pets during the research period. In some households, there were more monkeys than human beings. Seven species are represented: the red handed howler (*Alouatta belzebul*), the owl monkey (*Aotus infulatus*), the brown capuchin (*Cebus apella*), the Ka'apor capuchin (*Cebus kaapori*), the black-bearded saki (*Chiropotes satanas*), the squirrel monkey (*Saimiri sciureus*), and the golden-handed tamarin (*Saguinus midas*).

It should not be surprising that the Guajá would think that monkeys are most closely related to them. Elsewhere (Cormier 2002), I have characterized the Guajá as indigenous primatologists because their view is similar to the traditional goal of scientific primatologists in anthropology: to look to nonhuman primates in order to understand ourselves. They too recognize the close physical and behavioral similarities of nonhuman primates to humans and have attempted to account for this through their own cultural constructs.

The term "ethnoprimatology" was coined by Sponsel (1997). He identified six areas of study that such a focus of study would include: comparative ecology, predation ecology, symbiotic ecology, cultural ecology, ethnoecology, and conservation ecology. My work did not address all of the areas outlined by Sponsel, but I will briefly review them to give an idea of the potential scope of this type of research. Comparative ecology involves the study of ecological relationships between human and nonhuman primates. While primatologists tend to view humans as external to primate communities, ecological anthropologists tend to view nonhuman primates as part of the general landscape. Both miss opportunities to understand the ecological relationships shared by all primates. Predation ecology involves the study of nonhuman primate subsistence hunting by indigenous people. For primatologists, this involves increased attention to how human predation affects primate communities, and for cultural anthropologists, it involves looking at the contribution of nonhuman primates to human diets. Symbiotic ecology explores the interdependencies among human and nonhuman primate communities. Although human presence has often threatened nonhuman primate communities, not all human activities are destructive to them. Swidden agriculture, for example, creates secondary forest patches in which some primate species thrive. Agriculture may create a new food source for some primates. There are also monkeys living in some urban areas that subsist primarily on human provisions. Cultural ecology is a broad category that examines any cultural significance of nonhuman primates to human cultures. It could include such areas as food taboos, folklore, religion, mythology, pet keeping, and ornamentation. Ethnoecology includes indigenous perception, classification, knowledge, beliefs, and values regarding the nonhuman primates. And finally, conservation ecology investigates how human beings use and manage nonhuman primates as a resource. This would include both the positive and negative effects of human beings on primate communities.

The most critical area in which cultural anthropologists and primatologists can contribute to one another's disciplines is conservation. In ethnobiology, the notion of biocultural diversity is being increasingly employed. It involves the recognition that cultural survival and endemic species survival are not separate realms of inquiry, but mutually influence one another. It is particularly true in the tropical forests where cultures are rapidly changing and many primate species are threatened with extinction. As Fuentes and Wolfe (2002: 1) put it, "human and nonhuman primates share intertwined destinies."

CHAPTER OVERVIEW

This book explores the relationship between the Guajá and local monkeys through history, ecological adaptation, social organization, and cosmological beliefs. Chapter one provides a review of the history of the Guajá. Here, it is argued that the Guajá were members of a linguistic subgroup of Tupí-Guaraní speakers living as horticulturists in the 18th century. One or more of the myriad aftereffects of colonization, which included slavery, introduced diseases, and warfare likely led to their displacement, eastward migration, and loss of agriculture. Chapter two provides a review of the evolutionary history of the platyrrhines in South America and provides an overview of ecological relationships between human and nonhuman primates in Amazonia. Chapter three addresses the importance of monkeys in Guajá ecological adaptation. Monkeys provide a key game source for the Guajá in the wet season. Further, the Guajá ethnobotany reflects extensive knowledge of plants eaten by monkeys. It is also suggested that the anthropogenic forests the Guajá have traditionally exploited for babassu palms is an environment that may attract numerous monkey species. Chapter four discusses the Guajá kinship system as a preface for understanding the ways in which monkeys are incorporated into the Guajá kinship system. Guajá kinship terminology, marriage patterns, genealogical amnesia, and notions of plural paternity are discussed with attention to the importance of the siblingship and the role of affinity in their system. Chapter five discusses animism and life classification among the Guajá. Here, the Guajá use siblingship terms of affinity or partial consanguinity in classifying all forms of life. It also includes a discussion of Guajá notions of space and time. Chapter six discusses the role of pets in the Guajá community. While the Guajá share characteristics with other Amazonian groups, they differ from some in that pet keeping is more closely linked to the mother role within the group than to than to predatory relationships outside the group. The consequences for the monkeys of being kept in captivity by the Guajá are also addressed. Chapter seven discusses symbolic cannibalism as a trope for understanding Guajá relationships to nonhuman beings and the relationships among various forms of life in their natural and supernatural world. All consumptive relationships are understood as kinship relationships, and thus can be considered as conceptually cannibalistic. Chapter eight provides a discussion of the implications of ethnoprimatology beyond Amazonia, including a discussion of Western perceptions of primates. An appendix is also provided with a brief review of social and behavioral studies in primatology of the primate species found in the Guajá area.

Kinship with Monkeys

FIGURE 1.1 Trekking with Brown Capuchins *(Cebus apella)*

1

A BRIEF HISTORY OF THE GUAJÁ

The early history of the Guajá is not well known, but they probably originated on the lower Tocantins River and, like numerous other groups, migrated eastward in the wake of colonization. A number of events may have spurred their migration, including Portuguese slaving, introduced diseases, and warfare. One or more of these events was likely responsible for their shift from a horticultural to a foraging mode of production, including the possibility of their being actually enslaved by the Portuguese at one time. In their recent culture history, the Guajá remained largely isolated until the mid-1970s. Since that time, many of the same factors at work in their earlier history have affected them, including loss of territory, infectious disease, and conflicts with neo-Brazilians and other Amerindian groups.

EARLY HISTORY

The Guajá are members of the Tupí-Guaraní linguistic group, whose speakers are hypothesized to have lived to the west, in what is now the state of Rondonia, before dispersing through Amazonia. Rodrigues (personal communication in Jensen 1999) has suggested that the Tupí-Guaraní branch of the Tupí family spread from Rondonia in four waves of migration: 1) a southern wave that devel-

FIGURE 1.2   Black-bearded saki (*Chiropotes stananas satanas*)

oped into the Guaranían languages; 2) a wave east to Bolivia where Sirionó and Guarayu developed; 3) another eastern wave to the Atlantic Coast where Tupinambá developed; and 4) one or several waves to the north and east in the greater Amazonian regions. The Guajá are likely part of the waves described to the north and east. Rodrigues (1984/1985; see also Jensen 1999) classified the Guajá language in subgroup eight of Tupí-Guaraní, which is now considered to comprise eight additional languages, further divided into a group north of the Amazon (Emerillon, Wayampi, and Zo'é) and a group south of the Amazon which includes the Guajá, Anambé, Ka'apor, Takunyapé, Turiwára, and Amanayé.

Little is known about the early history of the Guajá. They have been living for at least the last 150 years in the vicinity of the Pindaré, the Turiaçu, and the Gurupi rivers in what is today the western portion of the state of Maranhão Brazil. The earliest definite description of the Guajá comes from a 1853 report by the president of the province of Maranhão where they were seen along the affluents of the Caru and Gurupi rivers (Gomes 1985b). They were seen again in 1873 by Dodt (1939) near the Gurupi River, and again reported by Amaral in the newspaper *O Maranhão* in 1895 (Jucá Filho 1987). Later, in 1913, a report of the now defunct SPI (Serviço de Proteção aos Índios) described a group of Guajá on the right bank of the Pindaré River (Gomes 1988).

Balée (1994a:25) has suggested that a group described by Noronha (1856), the "Uaya," living on the lower Tocantins in 1767 are probably antecedents of the Guajá who migrated eastward. Nimuendajú (1948:135) made the same connection, citing a 1774 mention of the Uaya along the lower Tocantins by Ribeiro de Sampaio. In addition, Nimuendajú believed that the Guajá were the same nomadic group as the "Ayaya," described as living between Belém and Imperatriz in 1861 and the "Uaiara" on the upper Gurupi in 1862. According to Nimuendajú (1948:135), the term "Guajá" is a derivation of an earlier Brazilian term for the Guajá, "gwazá" which itself likely derived from the term "wazaizara" (feather ornament owner) used by the Guajajara and the Tembé for the Guajá (Nimuendajú 1948:135).

Even if the Guajá were not the Uaya of the Tocantins, it is likely that they were previously part of the Amerindian Tocantins group. In addition to the Guajá, evidence suggests that four groups belonging to Tupí-Guaraní subgroup eight were probably living on the Tocantins during the middle and later eighteenth century, with three of them migrating eastward across the Moju, Acará, and Capim rivers. The Turiwára (now extinct) were among a number of groups described by Noronha (1856:8) as living among the along the Tocantins in 1767. Later Capuchin missionary reports placed the Turiwára to the east on the Acará margin in 1864, and farther to the east on the Capim in 1873 and 1875 (Arquivo

Público do Estado do Pará [APEP] 1864b, 1873:1184, 1875a). The Amanayé ("Amanajóz") were also described by Noronha along the Tocantins in 1767 and were described by Capuchin missionaries in the east along the Capim in 1873 and 1877 (APEP 1873:1185, 1877a).[1] The Amanayé are now considered nearly extinct, with only a few isolated families known to exist on the right bank of the Capim in the state of Pará (Correia de Alencar 1991). In addition, the Ka'apor have migrated eastward to their present location in the vicinity of the Guajá. According to Balée (1994a:25), ethnobotanical evidence suggests that the Ka'apor lived on the Tocantins at one time. Using oral histories and historical documents, he traces their eastward migration from the Acará (about 1790) to the Capim (1820s) to the Guamá and Piriá (1861) to the Gurupi (1872), near their present location (Balée 1994a:30–35). Capuchin missionary reports indicate that a Ka'apor group was still living in the vicinity of the Capim in 1873 (APEP 1873:1185), and about twenty were documented in the area of the Gurupi in 1874 (APEP 1874:1230). In addition, the Anambé were reported on the Tocantins in 1864 and in 1878 (APEP 1864a, APEP 1878). A small group of approximately seventy-five Anambé remain today in the vicinity of the Tocantins (Centro Ecumênico de Documentação e Informação/Projeto Estudo sobre Terras Indígenas no Brasil [CEDI/PETI] 1990:66).

An eastward migration also occurred among other Tupí-speaking groups. The Tembé were reported on the Acará in 1871, the Capim in 1873 and 1877, and on the Gurupi in 1887 (APEP 1871, 1873:1184, 1877b, 1887). Also, the Guajajara were reported along the Capim in 1873 (APEP 1873:1185) and today live in the vicinity of the Guajá and the Ka'apor. The Timbira, who are not Tupí-speakers, were another group described by Noronha on the Tocantins in 1767 and later reported along the Capim by Capuchin missionaries in 1873 (APEP 1873:1185).

One of the most notable features of the Guajá culture is that mid-nineteenth-century reports describe them as living as hunter-gatherers with no agriculture. Although they are linguistically related to numerous groups in the area and live in a similar habitat, their mode of production differs. Balée (e.g., Balée 1992, 1994a, and 1999) has argued that the Guajá were at one time horticulturalists who lost agriculture due to the effects of colonization, warfare, and epidemic disease. Guajá ecological adaptation will be addressed at length in chapter three, but the events of the early colonial period that may have led to a change in their mode of production will be reviewed here.

Between 1615 and 1616, the Portuguese established two settlements on the eastern coast of Brazil, leading to the creation of the State of Grão Pará e Maranhão in 1621 and the eventual displacement of the remaining French, Dutch, and English settlers by 1643 (Sweet 1975:45–47). Missionary groups also began entering the area during this time, with the arrival of the Franciscans in 1617, followed by the Carmelites in 1626, and the Jesuits (Companhia de Jesus) in 1636 (Sobral 1986:3; Cruz 1963 180–82; Lúcio de Azevedo 1930). From 1655 until they were expelled from Pará in 1757, the Jesuits set up mission *aldeias* (farming villages) using Amerindians who were enslaved or virtually enslaved (Lúciode Azevedo 1930; Sweet 1975 49–50).

Just before the expulsion of the Jesuits, the Companhia Geral do Comércio do Grão Pará e Maranhão was founded in 1755 (Cruz 1963: 43). Like the Jesuit missions, the Companhia operated through the forced labor and enslavement of Indians in the area, which was supported by law. The Lei de Liberdade dos Índios and the Promulgação do Diretória (1757) mandated compulsory labor of Indians between the ages of thirteen and sixty, requiring them to hunt, farm, cut wood, provide domestic services, construct public works, and collect forest products (*drogas de sertão*) (Brito 1991). Although considerable information exists in the Belém archives regarding Indian labor in the Companhia Grão Pará, it is difficult to extract specific cultural information because they were treated generically and given Portuguese names.

It is quite possible that the Guajá were a group who were at one time enslaved by the Portuguese, which may explain the change in their mode of production. Sweet's (1975) description of Portuguese slaving practices lends credence to the feasibility of this hypothesis. First, large groups of slaves were often captured at the same time. One recorded slaving expedition in 1653 brought back over five hundred Indians from the area of the Tocantins River (Sweet 1975:124–125). At times, whole villages were enslaved. In the early seventeenth century, one slaving practice involved erecting a cross in a village, leaving the area, and then returning a few months later; if the cross had fallen or deteriorated, the entire village and the group's descendants would be considered slaves (Sweet 1975:120). In the mid–seventeenth century *decimentos*, slavers, attempted to entice entire villages to resettle near Belém with the promise of European trade goods; however, if they refused, they were brought by force and, in effect, enslaved (Sweet 1975:125). Those who managed to escape, displaced far from their native villages and fields, sometimes resorted to plantation raiding to survive. An 1730 expedition to recover escaped Indian slaves returned over one hundred who had been living as a group and supporting themselves by raiding plantations for supplies (Sweet 1975:164).

It is also worth noting that the Guajá name for one of their divinities, *kapitã*, may suggest at least exposure to Portuguese slavery practices. The term is cognate with the Ka'apor term for headman (Balée 1994a:168) and probably derives from the Portuguese labor bosses, called *capitão*. The term may derive from Lingua Geral (Nheengatu), the Tupinambá based pidgin, which may or may not indicate direct interaction with the Companhia. Lingua Geral was in wide-spread use among the Portuguese in the seventeenth and eighteenth centuries and was the only language permitted among slaves; however, it also functioned as a form of intertribal communication among groups throughout the sur-rounding area (Brito 1991; Jensen 1999; Sweet 1975).

It should be mentioned that it would not have been necessary for the Guajá to have been actually enslaved for a similar effect to have occurred. In the mid-1650s, a Jesuit priest reported that a "hundred league" stretch of the Tocantins had been abandoned out of fear of Portuguese slavers (Sweet 1975:126). Al-though disease certainly was a factor, the Portuguese accounted for the depopu-lation due to their witnessing Indians disappearing into the forest rather than taking the risk of being taken off by slavers (Sweet 1975:122).

The effects of Old World infectious diseases on the Amerindian populations cannot be underestimated as a factor influencing migration and outright de-population. Numerous epidemics of smallpox spread though what is know Pará and Maranhão in the seventeenth and eighteenth centuries. One of the earliest documented was an epidemic of smallpox in 1621 that arrived at São Luis in Maranhão by means of a ship travelling from Pernambuco (Sweet 1975:79). Ad-ditional epidemic outbreaks of small pox were documented in the region in 1644, 1663, 1673, the mid-1690s, 1724–1725, 1776, and 1779 (Bettendorf 1910; Fonesca de Castro 1996; Sweet 1975). Fonesca de Castro describes the Indians' fleeing from their work posts during the 1776 and 1779 epidemics under the Companhia do Grão Pará e Maranhão.

Due to both the high mortality rate and the flight of Amerindians in re-sponse to epidemic disease, increasing numbers of African slaves were brought into the region (Sweet 1975). The danger was then compounded for Amerindi-ans with the additional threat of Old World tropical diseases. Yellow fever is be-lieved to have first arrived in the New World in 1648 (Yucatan and Havana) and malaria is believed to have arrived in the Amazon some time after 1676 (McNeill 1976). Cholera was also likely a problem in the area in the eighteenth century; numerous outbreaks were occurring in Europe and other areas of the Old World, including an outbreak in Portugal in 1833 (Marks and Beatty 1976). In the mid-nineteenth century, there were reports by the explorer Herndon (2000 [1851–1852]) of an outbreak of small pox (in 1849) and yellow fever (in 1850) in Pará. According to Cleary (1998), this was the first instance of yellow fever in the area. By Herdon's estimate, 25 percent of the Indians died as a result of the twin epidemics. It is worth noting that the first documentation of the Guajá in Maranhão occurs in 1853, only a few years after these outbreaks.

Another important event that may have spurred Guajá migration and change in mode of production was the Cabanagem civil war of 1835–1841. It began as a conflict between Portuguese loyalists to the crown's administration in Rio and native-born Brazilian elites seeking independent governance in Grão Pará; however, the rebels came to include escaped African slaves, the largely detribalized Amerindian population in the area, and those of mixed ethnicity (Cleary 1998). Forline (1997) suggests the Cabanagem war may have been the key event responsible for the dispersal of the Guajá. It was a devastating event for the region with an estimated 20,000 to 30,000 people killed (Cleary 1998). The first documentation of the Guajá in Maranhão, from the decade following the Cabanagem, also lends weight to Forline's view, but the Guajá migration may have been due to any combination of flight from epidemic disease, escape from Portuguese slavers, and warfare (including internal Amerindian warfare).

Displacement alone seems insufficient to explain the Guajá's loss of horticulture. Even if displaced, one would think they would have been able to reestablish themselves as at least part-time horticulturalists, particularly with a crop with low start-up costs such as maize (Balée 1992, 1994a). Further, displacement cannot explain their losing the critical skill of fire-making. Their continued nomadism would have to be explained in terms of displacement from their territory and either an inability to establish new territory for horticulture or losing their knowledge of how to propagate crops. An inability to establish new territory could be explained by depopulation and weakness in competition with larger Amerindian groups. Over several generations, this could have led to a loss of horticultural knowledge. According to Balée's (1988, 1994a) oral history information from the Ka'apor (current neighbors of the Guajá), they would not have allowed the Guajá to maintain permanent settlements. Alternatively, the loss of knowledge could have resulted from actual enslavement of a generation of Guajá, which might also explain their loss of indigenous fire-making technology.

One additional possibility is that the Guajá mode of production is either long-standing, existing prior to colonization, or that adoption of a foraging lifestyle may have involved an element of choice. Erikson (in press) describes one consequence of Matis (Panoan-speakers) depopulation from epidemic disease; they have been *able* to incorporate more game meat into their diet. In other words, the reduction in population has allowed a larger per capita consumption of game meat in the area and allows for a reduction in the reliance on agricultural crops in the diet. Erikson's description suggests a choice in the degree of reliance on hunting instead of agriculture, and it echoes arguments made by Rival (1998) that the limited use of agriculture among the Huaorani of Ecuador is a political choice critical to their group identity on multiple levels, rather than reflecting cultural loss or horticultural devolution.

The possibility that foraging may be adopted by a group because it is desired or possible, and not because agriculture is impossible, is certainly worthy of careful consideration. Rival (1998:241) states that "Huaorani ethnography reminds us that the food quest is embedded in sociocultural processes, and that the 'regressed agriculturalists' of Amazonia, in giving up more intensive forms of agriculture have exercised a *political choice*." I am in full agreement with Rival on the point that the quest for food is embedded in sociocultural processes. Much of this work is aimed at elaborating that very point in the case of the Guajá. A group's ecological adaptation is not merely a component of their way of life, but is well integrated into social and symbolic dimensions of their culture.

However, I do believe that a critical divide exists between horticultural groups on a continuum from limited to intensive agriculture and groups living among horticulturalists who have no knowledge of agriculture at all. As will be discussed in the next section of this chapter, even "indigenous foragers" who do not plant in the tropical forests typically incorporate agricultural crops in their diet to some degree through trade with food producers. And perhaps more importantly for the specific case of the Guajá, it is difficult to imagine a scenario wherein they would make a political choice to abandon fire-making technology in favor of carrying firebrands. Although I do not think that foraging represents an actual political choice for the Guajá, it most certainly has political meaning and political consequences, both in terms of interethnic politics in the region and in their own ambivalence over the recent introduction of agriculture to their society (see Cormier in press).

THE "WILD YAM QUESTION"

The Guajá are somewhat anomalous for tropical forest hunter-gatherers because they are not involved in trade relations with agriculturalists. Balée (1989, 1994a) has argued that the Guajá rely heavily on anthropogenic environments, utilizing species growing in the old fallows of horticultural peoples. This ties into what has been called the "wild yam question," referring to Headland's 1987 article by the same name. As will be discussed below, the argument takes several forms, but in essence it suggests that there are no true hunter-gatherers in the tropical rain forests because wild plant foods are inadequate in that environment to sustain them. Thus, these foraging peoples must depend on agriculturalists in some way for domesticated plant foods.

One argument against the possibility of independent hunter-gatherers describes the rain forest flora as lacking adequate nutrients for human beings. Despite the great biodiversity of the tropical rain forest, it has been argued that this habitat is difficult for humans to live in. First, most of the edible vegetation in the rain forest is located high in the canopy; much of the remaining vegetation contains toxins that render it inedible for human beings (Hutterer 1983). The mostly terrestrial African rain forest gorillas overcome these problems with an enlarged colon and slower digestive processes that allow for detoxification and fermentation of the cellulose in leaves (Fleagle 1999). For human beings, the great biodiversity of the plant species is disadvantageous because edible species may be widely dispersed and thinly distributed (Hutterer 1983). Further, because plants do not face major seasonal stress, very little energy is concentrated in nuts and seeds that are rich in fats and complex carbohydrates, with the exception of palms (Hutterer 1983).

Many researchers have stressed the lack of carbohydrates as the critical limiting factor for hunter-gatherers, although Bailey et al. (1989) include fats and other nutrients as important components (also see Hart and Hart 1986). Headland (1987) has stressed the scarcity of complex carbohydrates. He argues that for the Agta of the Philippines foods such as wild yams, palm piths, and other tubers are too scarce in the rain forest to meet their carbohydrate needs and that honey, although important, is only seasonally available. Headland further argues that carbohydrate scarcity was a critical limiting factor for prehistoric peoples' movement into tropical rain forests and that hunter-gatherers can only live in the rain forest if agriculturalists are present.

Milton (1984) also views lack of carbohydrates as the critical problem for the Maku hunter-gatherers of the Northwest Amazon. She argues that although enough game is available to meet protein requirements, they lack sufficient carbohydrates in their diet. This may seem to run counter to Harris's (1974) well-known argument that protein scarcity in the Amazon leads to the warfare/female infanticide complex. This controversial thesis does not apply to hunter-gatherers. Rather, the argument rests on the heavy dependence of the agricultural Yanomamö on protein-poor bananas and plantains, combined with local depletions of game around the settled village.

Relying on protein as a source of calories presents its own problems, as Speth and Spielmann (1983) have explained. Lean meat is high in protein, but relatively low in calories. Heavy reliance on lean meat increases metabolic rate and increases the number of calories needed each day. Further, carbohydrates, and to a lesser extent fats, have a protein-sparing action. Therefore, eating a high protein diet increases the amount of calories needed and increases the chances that one's own body protein will be broken down for energy. Carbohydrates have the strongest protein-sparing action. However, Speth and Spielmann point out that lack of fats can lead to a shortage of fat soluble vitamins and fatty acids, such as linoleic acid. Hart and Hart (1986) found that among the Mbuti, during the five months of the year when wild plant foods were scarcest, game meat was also at its leanest.

Many researchers have argued against the possibility of independent hunter-gatherers in the rain forest on the basis of evidence of trade relations between foragers and food producers. In the Philippine rain forest, the hunter-gatherer Agta people trade wild game and other forest products for rice and other agricultural products (Headland 1987). In the Amazonian rain forest, the Maku hunter-gatherers trade wild game, baskets, and labor for manioc, other cultigens, and Western goods (Jackson 1983, Milton 1984). Jackson has gone so far as to describe this as a master-servant relationship. Congo Basin foragers (the so called "pygmies") trade forest products, such as game and honey, to the Bantu and Sudanic-speaking horticulturists who give them cassava, rice, bananas, and other crop foods (Bailey et al. 1989). Bailey and Peacock (1988) argue that none of the foraging peoples in the Congo Basin are known to live for more than a few weeks without relying on agricultural food. The Mbuti of the Ituri forest trade labor and forest products for iron tools and agricultural products, including cassava, plantains, cultivated yams, and sweet potatoes (Hart and Hart 1986:31). The Efe of the Ituri forest trade wild game, honey, fruit, and building materials in exchange for cultivated food, cloth, and iron tools of the agricultural Walese (Bailey and Peacock 1988). Further, Bailey and Peacock (1988) found that 63.5 percent of the calories of the Efe come from Walese agricultural foods.

An important criticism of the impossibility of tropical forest foraging is that riverine resources have been ignored. Brosius argues that some of the most productive varieties of wild yams in Southeast Asia exist in riverine areas (Burkhill 1954:293–437, cited by Brosius 1991:133). Brosius (1991) argues that upland hunter-gatherers today may have been able to live independently in the past near rivers. Related to this is the criticism that fish have been ignored as a productive resource for hunter-gatherers (e.g., Beckerman 1983, Colinvaux and Bush 1991, Gragson 1992). This is further supported by recent evidence from the Amazon rain forest of "riverine foragers" with pottery, living approximately 7000 to 8000 years ago (Roosevelt et al. 1991). Bailey and Headland (1991) have taken note of this argument and agree that fish may be a key resource for hunter-gatherers, especially during the dry season when fish are easier to catch, which is the very time when plants are scarcest and game animals leanest.

In Amazonian anthropology, a long-standing ecological dichotomy has existed between riverine versus upland habitats. Meggers (1957, 1973) stressed the importance of the greater agricultural potential of the alluvial soils on the floodplains versus the relatively infertile soils of the upland forest. Carneiro (1964)

stressed the greater protein potential of aquatic resources versus the more readily depleted game in the upland areas. Lathrap (1968) agreed with Carneiro and
expanded his riverine-versus-upland distinction into a historical explanation for
the current distribution of groups in the Amazon Basin. He proposed that
groups living in the upland areas do so now because they have been historically
marginalized there. Increasing population density at the riverine sites would
create the need to move farther and farther from these sites in order to make a
living. Gross (1975) also has proposed that archaeological evidence may reveal
this essential distinction between the riverine and upland peoples. He argued
that upland groups may have been dependent on riverine groups for protein resources and may have been subordinate to these riverine groups. Moran has
criticized the riverine/upland dichotomy as distorted and overly simplistic, for it
fails to take into account the heterogeneity of soils and ecosystems in the Amazon Basin, especially in the uplands (Moran 1990, 1993). While his criticism is
important, the increased availability of fish in some riverine areas (particularly
whitewater) does create greater potential for higher population densities, as he
notes himself (Moran 1993: 89).

A second important area of research on tropical forest foraging explores
hunter-gatherers' indirect reliance on food producers through their exploitation
of plants that grow in the old fallows of agriculturalists. Bailey et al. (1989) do
take this into consideration, commenting that commonly exploited plants, such
as nuts and yams, are light-dependent, and that swidden ("slash and burn") agriculture has allowed these plants to exist in greater densities. Prior to swidden
agriculture, these plants existed mainly along streams and in the light-gaps of
treefalls. They argue further that swidden agriculture has created ecotones
within the tropical forest. They suggest that Central African foragers lived on
the savanna–forest ecotone until shifting cultivators arrived in the area and created similar savanna–forest ecotones within the forest.

In addition, many species of palms grow in old fallows (Balée 1988, 1989),
and several researchers have discussed the importance of reliance on palms,
which indicates an indirect dependence on the presence of agriculturalists. According to Brosius (1991), a staple source of carbohydrates of the Penan hunter-
gatherers is the sago palm (*Eugeissona utilis*), at least some of which are exploited in the old fallows of agriculturalists. Townsend (1990) has described the
Sainyo-Hiyowe hunter-gatherers as also relying on a sago palm (*Metroxylon*)
that they cultivate with approximately 10 percent of their diet coming from domesticated crops.

In sum, the ethnographic evidence seems to support the proposition that
modern hunter-gatherers are not living in the *upland* tropical forest independently of food producers. The dependence on food producers, however, may be

direct or indirect. Hunter-gatherers either trade with food producers or rely heavily on plants (particularly palms) that grow in old fallows.

Evidence suggests that the Guajá have not been independent hunter-gatherers either. Early reports by Beghin (1957) and Nimuendajú (1948) describe their stealing food from the fields of neighboring Indians and Brazilian settlers. However, sporadic crop-stealing was unlikely to have affected their diet in a major way. More importantly, Balée (1988) has shown the Guajá relying heavily on the babassu palm (*Orbignya phalerata* = *Attalea speciosa*) which grows in the abandoned fields of nearby food producers. Balée (1994a) has gone so far as to classify the babassu (and other old fallow colonizers) as a semi-domesticate because it thrives in the wake of domestication.

Nonetheless, the term hunter-gatherer is probably the most accurate classification for the Guajá, not only because they continue to spend considerable time as nomadic foragers, but also due to other features of their culture that are commonly found among hunter-gatherers, including their family band social organization, their basically egalitarian social relations, and their demand-sharing economy, all of which will be discussed in later chapters.

RECENT HISTORY

Today, the Guajá are located on the western edge of state of Maranhão, Brazil, in a region called pre-Amazonia. The region is similar to Amazonia in terms of tree species, geology, and climate, but its major rivers do not drain directly into the Amazon River (see Balée 1994a: 9–12). Although they were a foraging people when first encountered, many of the Guajá are now undergoing a transition to a horticultural mode of production under the auspices of the FUNAI. Guajá families only began coming into contact with non-Indians in 1973, and some still remain mostly or completely isolated. They live in four indigenous areas overseen by posts established by the FUNAI.

All research activities were conducted in the vicinity of the P.I. (Posto Indígena) Awá on the Caru reserve, which has the largest number of Guajá in contact, a population of 108 individuals at the time of this study (see Figure 1.1). The Caru reserve, bordered by the Pindaré River to the east, is approximately 172,000 hectares in area (Gomes 1988) and is shared with the Guajajara Indians. To the north of the Caru is the Alto Turiaçu Reserve, which is approximately 530,500 hectares (Balée 1994b) and is shared with the Ka'apor and some Tembé Indians. There are also Guajá in contact in the Awá Indigenous area, which is approximately 240,000 hectares (Gomes 1991) and situated in between the western sides of the Caru and the Alto Turiaçu. In addition, there are a little-known group of Guajá living to the south of these reserves in the Araribóia indigenous area, also shared with the Guajajara.

In all, there are approximately two hundred Guajá in contact today. In addition, some Guajá are known but uncontacted or infrequently seen, including some fifteen to twenty on the Alto Turiaçu Reserve, about twenty on the Caru reserve, and about thirty to forty on the Araribóia indigenous area (Damasceno da Silva, 1997, personal communication; Gomes 1996). Thus, the total population of the Guajá is probably around 265, or more likely between 250 and 275, with about two-thirds to three-fourths of the population now in contact.

The first consistent contact with the Guajá Indians occurred in the context of illegal development of their habitat, which has continued unabated to this day.[2] The first invasion of the Guajá area in Maranhão can be traced to 1958, when a man named Carlos Gomes moved into the region of the Pindaré, although the area was supposed to be under the protection of the SPI (F. Parise 1987). By 1969, approximately forty families had invaded the area along the Pindaré (F. Parise 1987). That same year the BR-222 was built (Gomes 1988), a roadway that connects São Luis with Imperatriz, and this cleared the way for more infiltration of the area. Around the same time, the FUNAI began attempting to make contact with the Guajá. A number of sporadic sightings were made until 1973, when official contact was made. These early reports characterize the Guajá as living in small, nomadic family bands of ten to twenty individuals and relying on babassu palms, howlers, and tortoises as major subsistence foods. (e.g., Beghin 1951, 1957; Dodt 1939; Carvalho 1992; Meirelles 1973; Nobre de Madeiro 1990; Gomes 1988; Nimuendajú 1948; V. Parise 1988). Nimuendajú (1948) received his information from reports by the Tembé in 1913–1914 and the Guajajara in 1929.

The consequences of recent contact have been devastating to the Guajá, with countless deaths through disease and outright murder along with the confiscation and destruction of their habitat. In some instances, organized posses set out to kill the Guajá (Gomes 1988). The greatest threat has been infectious disease. Numerous reports exist of the Guajá raiding the homes of the early illegal invaders (see V. Parise 1973b), providing ample opportunity for exposure to new diseases. By 1971, the Brazilians in the area were reporting large numbers of Indian corpses in the forest on the left and right banks of the Pindaré River (V. Parise 1973b). FUNAI reports describe a harrowing scene in 1971 after coming into contact with three Guajá children. The children took FUNAI officials to their camp where their mother had apparently died in childbirth, and the body of their father lay dead and decomposing nearby (V. Parise 1973b). A similar incident occurred in March of 1973 when a family of fifteen was contacted by FUNAI. By August, all were dead except for two children, the elder of whom later died of tuberculosis (F. Parise 1988; V. Parise 1973a, 1973b).

Subsequently, the P.I.(Posto Indígena) Guajá was created on the Alto Turi-açu Reserve. By 1976, ninety-one Indians were in contact at the P.I. Guajá; how-ever, after a respiratory virus was introduced, over two-thirds died, leaving a pop-ulation of only twenty-eight Indians by 1981 (F. Parise 1988). Less dramatic, but still clearly devastating, was the loss of 20 percent of another group of Guajá: in 1980, after the P.I. Awá was created on the Caru Reserve, a group of twenty-eight Guajá were moved there by the FUNAI to protect them from the threats of in-vaders in their habitat. Six died from illness before transfer was completed (Gomes 1996).

One of the most important events in the recent history of the Guajá was the construction of the state owned Companhia Vale do Rio Doce (CVRD) railway for the Ferro Carajás mining project in 1985. The CVRD (supported in part by the World Bank) has been responsible for greatly facilitating illegal settlements in the region and has also been directly responsible for massive deforestation. According to Balée (1994b:2), approximately 2,350 square miles of forest per year have been cut down and burned for use in pig-iron smelting factories. In eastern Amazonia in general, more than half of the original forest has been cleared for agriculture and development (Lopes and Ferrari 2000).

In the first year of construction, three incidents occurred involving Guajá shooting and wounding Brazilians with arrows, two of them in the vicinity of the railway construction (Gomes 1985a). The Indians presumably responsible for the railway shootings were transferred to the Caru Indigenous Reserve (F. Parise 1988). Although these Indians spoke the Guajá language, the Caru Guajá did not recognize them as being true Guajá (*awa-te*) and designated them as "*mihúə*."

Thus, these indigenous reserves are artificial to some extent. The Caru re-serve, for example, includes several groups whose original range was in the re-gion and two other groups transferred to the reserve. In addition, three individ-uals have been transferred to the reserve whose families had died. Two were survivors of a massacre by local Brazilians. One of the two, Karipiru, underwent an incredible saga described by Toral (1991). When his group was attacked, he was shot in the back, his young son was captured, and the rest of the group was killed or scattered. Karipiru lived alone in the forest, while his son was taken to FUNAI in São Luis, where he was then raised as an interpreter. Ten years after the attack on his group, Karipiru came into contact with a group of construc-tion workers in Bahía, some 600 km. from his original home. When local Indi-ans were unable to understand the language he spoke, FUNAI brought in an in-terpreter to try and determine what group he was from. When the interpreter saw Karipiru, he asked him to lift up his shirt, and when he saw the scars on his back from the gunshot wounds, he realized it was his own father. Karipiru now lives on the Caru reserve. In 1987, another young man from the same group was found who had wandered even further to Minas Gerais (F. Parise 1987). This is the same young man described previously who had been involved in the murder that occurred while I was there.

Although the major threats to the Guajá land stem from invasion and the development of their habitat by non-Indian Brazilians, tension also exists within the reserves for land. On the Caru reserve, the Guajajara, old enemies of the Guajá, share the reserve and also have been encroaching on the habitat of the Guajá. While the Guajá have maintained much of their traditional lifestyle, many Guajajara are more acculturated. According to the local FUNAI, the Guajajara have been illegally renting land to neo-Brazilians who have cleared it to grow rice. In addition, the Guajá on the Caru reserve have hostile relations with another uncontacted foraging group they call the "*Aramaku*," who are probably distantly related Guajá. One Guajá man was wounded by an arrow shot by the *Aramaku* during the research period. Although he was not seriously injured, he fired back with a shotgun and believes he seriously injured one of them.

The threats to the Guajá habitat are also threats to the endemic species within their habitat on which they have traditionally relied, including Neotropical monkeys. The ecological role of primates in the diet of the Guajá is the subject of chapter four. Prior to that discussion, the following chapter will present an overview of the history of human and nonhuman primate ecological relationships in Amazonia.

2

A BRIEF HISTORY OF NEW WORLD MONKEYS

Primatological research typically excludes the human primates, despite the fact that, as with all primate communities, mutual ecological influences exist. Describing ecological influences as mutual is not meant to suggest that the influences are equal. Clearly, the last five hundred years have brought increasing devastation of nonhuman primate communities by their human cousins. In this chapter, the history of human-nonhuman primate interactions in Amazonia are addressed. Emphasis is given to human behaviors which negatively affect primate habitats, such as habitat destruction, hunting pressure, and primate trade. However, additional ecological relationships are also included such as nonhuman-primate adaptation to anthropogenic forests and the role of primates in seed dispersal of edible plants for humans.

Important ecological relationships among primate communities can be seen as far back as the Miocene, when a decrease in the relative abundance of early apes coincided with an increase in relative abundance of cercopithecoid monkeys (see Fleagle 1988). The monkeys either filled a vacant niche left by the decline of the Miocene apes, or they may have directly outcompeted and replaced many early apes. In either scenario, the cercopithecoid adaptive radiation must be understood as contingent on the fate of the early apes, which may have implications for human evolution.

FIGURE 2.1 The Ka'apor capuchin (*Cebus kaapori*)

In Africa, hominid coexistence with other primate species extends back approximately five million years. However, current archaeological evidence cannot reliably place humans in Amazonia any earlier than 11,000 years ago (Roosevelt et al. 1991, Roosevelt et al 1996). Sponsel (1997) has argued that this is sufficient time for coevolution of human and nonhuman primate species. While large scale genetic changes may not be observable among the Amazonian primates within this short time span, there is clear evidence of phenotypic changes as nonhuman primates adapted or, in some cases, failed to adapt to human presence.

Neglecting human presence as an influencing factor severely distorts our understanding of the Neotropical primates. Many such interactions are amenable to historical ecology, an approach that focuses on the dynamics of human activity in an environment, rather than viewing humans as separate from nature (see Balée 1994a, 1998). Equally inadequate are views of nonhuman primates adapting to static ecosystems where humans are invisible. Here, overlap exists between the aims of ethnoprimatology and historical ecology, particularly in an examination of the exploitation of anthropogenic habitats by the Guajá and nonhuman primates.

AN OVERVIEW OF PLATYRRHINE ORIGINS, THE FOSSIL RECORD, AND SYSTEMATICS

There are sixteen genera of living platyrrhines, or New World monkeys. Traditionally, they were divided into two groups, the callitrichids, which included the marmosets and the tamarins, and the cebids, a category including all others except Goeldi's monkey (*Callimico*), which was judged to share affinities with both the callitrichids and the cebids. In addition, there has been broad consensus that there are three monophyletic groups of platyrrhines. The callitrichines are small bodied, have some clawlike nails, and include the marmosets species (*Callithrix* and *Cebuella)* and the tamarin species (*Saguinus* and *Leontopithecus*). The atelines are relatively large bodied with prehensile tails and include the woolly monkeys (*Lagothrix*), the muriquis or woolly spider monkeys (*Brachyteles*), the spider monkeys (*Ateles*), and the howlers (*Alouatta*). The pitheciines are specialized seed predators and include the sakis (*Pithecia*), the bearded sakis (*Chiropotes*) and the uakaris (*Cacajao*). The relationship of *Callimico* (Goeldi's monkeys), *Aotus* (the owl monkeys), *Cebus* (the capuchins), and *Saimiri* (the squirrel monkeys) to these three groups is less clear.

Recent DNA evidence has shed further light on the relationship of the "out-lying" monkeys. Schneider and Rosenberger (1996; see also Porter et al. 1997; Porter et al. 1999) have proposed three clades: the atelids/atelines, the pitheci-ids/pitheciines, and the cebids. The atelids remain unchanged in light of the DNA evidence, comprising the spider monkeys, woolly monkeys, muriquis, and howlers. The pitheciid group has been revised to include the titi monkey. This finding was a surprise, since it had long been assumed that owl monkeys and titi monkeys were closely related. Both are small-bodied frugivores with a monoga-mous social organization and morphological similarities in both their cranial and postcranial anatomy (Fleagle 1988). The cebid clade includes two mono-phyletic groups: one is the callitrichines and the other comprises both *Cebus* and *Saimiri* species. The status of *Aotus* has not been completely resolved among these researchers. While Schneider's DNA evidence links *Aotus* to the cebid group, Rosenberger believes that this grouping is inconsistent with *Aotus* morphology. Wright (1996), who has done extensive fieldwork on owl monkeys, suggests that part of the difficulty in classifying *Aotus* may be due to its retention of many primitive traits.

Platyrrhine Origins

The origin of the platyrrhine monkeys has been somewhat puzzling for prima-tologists. The first fossil platyrrhines appear in the fossil record in the late Oligocene, approximate 30 million years ago. Anthropoid primates, which in-clude all monkeys, apes, and human beings, originated in the Old World, and entrance into the New World would involve crossing considerable water barri-ers. Several hypotheses have been advanced for the origin and possible route of platyrrhines to South and Central America, but the most viable (or as Fleagle [1999:447] puts it, "the least unlikely") explanation is that they rafted from Africa across the Atlantic on floating islands.

The New World monkeys and the caviomorph rodents both share physical similarities with earlier African forms and are believed either to have rafted or island-hopped across the late-Eocene Atlantic Ocean, which was much narrower then than it is today (e.g., Fleagle and Kay 1997; Hoffstetter 1980; Lavocat 1980; Sarich and Cronin 1980). Since prosimians were present in North America until the early Oligocene, some have examined the hypothesis that New World monkeys originated from a prosimian-like primate migrating from North America; however, this model would also necessitate crossing a water barrier, since South America was an island continent from some time before the middle Eocene until the Pliocene (e.g., Delson and Rosenberger 1980; Gingerich 1980; see Wood 1980 for arguments of caviomorph rodent migration from Middle America; see Ciochon and Chiarelli 1980 for a general review of arguments on the North American origin vs. the African origin model). In addition, the platyrrhines possess numerous anthropoid features in their cranium and dentition distinguishing them from prosimians (Fleagle 1999). Further, recent work by Masanaru et al. (2000) comparing early platyrrhine fossils with early anthropoid fossils in Fayum Egypt suggests that the platyrrhines had also undergone some diversification in Africa prior to their arrival in South America.

Fleagle (1999) also notes that Szalay offered the novel hypothesis that anthropoids originated in the New World and traveled across the Atlantic to Africa. However, Fleagle argues it is less likely than the Africa to Atlantic route because it lacks the support of evidence linking earlier African forms of mammals and South American forms. In addition, Fleagle (1999) notes that the recent discoveries of probable middle-Eocene anthropoids in Asia could suggest an Asian origin of platyrrhines (via the Bering Strait to North America to South America). However, he dismisses this due to the lack of any anthropoid forms in North America.

Houle (1999) has recently examined the evidence that anthropoid primates traveled to South America via Antarctica. These continents were connected until the late Eocene and fossil mammals from Antarctica have been discovered that are related to South American forms. In addition (see below), fossil primates were present in Southern Patagonia in the early and middle Miocene. While the connection between South America and Antarctica is clear, the problem involves the route to Antarctica. Houle argues that a route from Asia to Australia to Antarctica is unlikely because it would have necessitated crossing two water barriers. Although these land masses were closer than they are today, by 50 million years ago, Australia had separated from Southeast Asia, and by 42 million years ago, Australia and Antarctica were separated by at least 1000 kilometers. In addition, there is no fossil evidence of primates in Australia. Houle also argues that a route from Africa to Antarctica across the Indian ocean is unlikely because it would have necessitated crossing a very large water barrier of 2,600 kilometers, as opposed to the 1,400-kilometer route across the Atlantic directly to South America. In addition, Houle examines the problem of surviving on a floating island during a transatlantic journey. After reconstructing Eocene wind and water currents in the Atlantic, he believes that a floating island could make the journey in only seven to eleven days, which should be long enough for monkeys to survive.

The Fossil Record

Fleagle and Kay (1997:3) described the status of the platyrrhine fossil primate record thusly: "Until recently, a large shoe box could contain the primate fossils from all of South America and the Caribbean from the last 30 million years." Although the fossil record remains rather sparse, there have been numerous discoveries that are providing new insights into platyrrhine evolutionary history.

The earliest known platyrrhine fossils are from the Late Oligocene of Salla, Bolivia. While Takai et al. (2000) classify the Salla primates in a single taxon, *Branisella boliviana*, Rosenberger, Hartwig, and Wolff (1991) have differentiated some into a separate taxon, *Szalatavus attricuspis*. Takai et al. (2000) describe affinities in dentition between *Branisella* and *Proteopithecus*, recently described in the late Eocene of Fayum, Egypt (see Miller and Simons 1997, Simons 1997). While Takai et al. (2000) acknowledge that the relationship of *Branisella* to extant primates remains unclear, they do make the case that its dental morphology indicates that it is ancestral to the callitrichine primates. This suggests that the platyrrhine monkeys may have diversified in Africa before crossing the Atlantic to South America.

Several primate genera have been identified from the early and middle Miocene of Argentina and Chile. Researchers have had difficulty establishing clear links between these fossil primates and living primates. Fleagle (1999) offers several possible explanations. The Patagonian primates may represent a sister group to the living primates rather than being direct ancestors. It is also possible that either the primitive features or the generalized features, or both, of these earlier primates are responsible for the contradictory affinities with living groups. In addition, the Patagonian primates may have adaptations specific to the geographic locale. Bown and Larriestra (1990) have described the Pinturas Formation of Santa Cruz province, where many of these primates are found, as a forested habitat with a climate that fluctuated between periods of wetness and desiccation.

Dolichocebus gaimanensis from the early Miocene of Argentina has been suggested to be linked to either *Neosaimiri-Saimiri* (Rosenberger et al. 1990) or to be a sister clade to all living primates (Fleagle and Kay 1997). The relationship of *Tremacebus harringtoni* (early Miocene, Argentina) to other Neotropical primates is also unclear, but cranial similarities to *Aotus* and *Callicebus* have been demonstrated (Fleagle and Kay 1997). The affinities of *Chilicebus carrascoensis* of the early Miocene of Chile are also unclear (Flynn and Wyss 1998,

Fleagle and Kay 1997), although some similarity to *Saimiri* has been described in its dentition.

Two primate genera have been recovered from the Pinturas Formation in Southern Argentina (17.5–16.5 M.Y.B.P.). The dentition of *Carlocebus intermedius* and *C. carmenensis* is most similar to that of *Callicebus* (Fleagle 1999; Fleagle and Kay 1997). *Soriacebus ameghinorum* and *S. adrianae* might demonstrate some similarity to both the pitheciines and the callitrichines, but it has been argued that the similarities to the callitrichines are convergent rather than derived (Kay 1990; Rosenberger, Setoguchi, and Shigehara 1990).

Homunculus patagonicus of the early-middle Miocene of the Santa Cruz formation on the Atlantic Coast of Southern Argentina has limb elements resembling callitrichids, but its dental affinities are more similar to *Aotus*, *Callicebus*, or *Alouatta* (Fleagle 1999). The recently described *Proteropithicus neuquenensis*, the youngest of the Argentine primates, has been classified as a pitheciine primate with dental affinities to *Soriacebus* (Kay, Johnson, and Meldrum 1998, 1999). This further supports the link between *Soriacebus* and the pitheciines.

A number of fossil primates have been described from the late Miocene of La Venta, Columbia. Paleoecological analyses describe the rainfall level of this environment as associated with a riperian mosaic forest, or transitional between savanna and forest environments, rather than undisturbed continuous canopy (Kay and Madden 1997). Unlike the Patagonian fossils, many of these show clear links to living genera. These date to approximately 11.6–13.5 M.Y.B.P. (Flynn and Wyss 1998).

Nuciruptor ribricae and *Cebupithecia sarmientoi* have been linked to the pitheciines (Fleagle 1999). Kay (1990) believes that *Cebupithecia* is a sister taxon to the living pitheciines rather than an ancestor. Kay describes a lack of some of the specializations of the pitheciines yet indicates similarities to the pitheciines in both the dentition and postcranial features. Two fossil species have been linked to *Alouatta* (howlers): *Stirtonia victoriae* and *S. tatacoensis* (Fleagle 1999). Although the former was apparently more robust, a number of analyses have clearly linked *Neosaimiri fieldsi* to the extant *Saimiri* (e.g. Nakatsukasa, Takai, and Setoguchi 1997, Rosenberger, Setoguchi, and Hartwig 1991, Takai 1994). Rosenberger, Setoguchi, and Hartwig described a fossil platyrrhine called *Laventiana annectens* as a primitive form of *Neosaimiri*, but Takai (1994) argues that further analysis of more than two hundred new specimens of *Neosaimiri* suggests that *Laventiana* should be included within the *Neosaimiri* taxon (and also refutes previous suggestions that *Neosaimiri* has a close relationship to *Branisella* or the callitrichines).

Somewhat less clear is the relationship between *Aotus dindensis* and *Mohanimico herskovitzi*. While *Aotus dindensis* has been classified as a fossil owl monkey (Fleagle 1999), Kay (1990) has argued that *Mohanimico* and *Aotus dindensis* are the same species and a sister taxon to the living owl monkeys. Rosenberger, Setoguchi, and Shigehara (1990) argue that *Mohanimico* appears to be a callitrichine, most similar to *Callimico* by its dental morphology. Three additional species have been linked to the callitrichines: *Microdon kiotensis* is known from only three teeth, but one study suggested that it is a fossil callitrichine (Rosenberger, Setoguchi, and Shigehara 1990). *Lagonimico condutatus* is large for a callitrichine but has been linked to them, and *Patasola magdalena* demonstrates some features similar to *Saimiri* and the callitrichines (Fleagle 1999).

Fossil primates from the Pleistocene-Recent deposits have been found in Brazil and the Caribbean. Recently, two new primate species have been described in Bahia, Brazil, dating to approximately 10,000 B.P., which are remarkably large for known platyrrhines (Cartelle and Hartwig 1996a, Cartelle and Hartwig 1996b, Hartwig 1995, and Hartwig and Cartelle 1996). *Protopithecus brasilensis* weighed approximately 25 kg (more than twice the size of any living platyrrhines) and is described as similar to the extant *Alouatta* (howler) in its cranial morphology and to *Ateles* and particularly *Brachyteles* in its postcranial skeleton. The second, *Caipora bambuiorum* resembles the spider monkey in its cranial and postcranial anatomy and weighed approximately 20 kg. Heymann (1998) has argued that because of both their large body size and because their habitat was likely cerrado (semi-deciduous with open woodland and some elements of unforested habitat), these primates may have been at least partially terrestrial.

A number of fossils have also been found in Caribbean sites in the late Pleistocene-Recent sites. *Xenothrix mcgregori* from Jamaica is an unusual primate that is difficult to link with living species. Its postcranial skeleton suggests slow, quadrupedal motion, which differs from the locomotion pattern of other platyrrhines (Fleagle 1999). Its dental formula is the same as that of the callitrichids; Ford (1986) suggested at one time that it could be an enlarged callitrichid. However, its molar morphology is now viewed as distinctive from the callitrichids (Fleagle 1999) and Rosenberger, Setoguchi, and Shigehara (1990) believe that the dental evidence suggests it is a pitheciine most closely related to *Callicebus*.

A partial cranium originally designated *Paralouatta varonai* from the late Quaternary in Cuba was initially thought to be related to *Alouatta* (Rivero and Arrendondo 1991). However, it is no longer thought to be related to howlers and may be related to *Callicebus* (Fleagle 1999). The dental remains of another Cuban primate, *Ateles anthropomorphus* (*Montaneia anthropomorphus*), was found with pre-Columbian human remains; however, it has now been dated to between 1670 and the present, suggesting that it was intrusive and the result of later primate trade (McPhee and Rivero 1996). It has been posited that *Antil-*

lothrix "*Saimiri*" *bernensis* from Haiti and the Dominican Republic has dental traits similar to squirrel monkeys (Ford 1990), but it has also been linked in the past to *Cebus* (Fleagle 1988). Its relationship to modern primates is unclear, but its dental morphology suggests a diet of hard fruit or seeds (Fleagle 1999).

To summarize, little has been known about Neotropical primate evolution until recently. Numerous new fossils have been recovered, which has coincided with the emergence of new technologies in molecular biology. Some differences exist between the DNA evidence and the morphological evidence, and there are differences as well among those researchers analyzing the fossils. The general picture is that Neotropical primates probably arrived from Africa to the South American continent in the late Oligocene, approximately 30 million years ago. The fossils from the late Oligocene through the middle Miocene have been difficult to link with living genera. This difficulty may be due to their retention of primitive features, the fossils' belonging to sister clades rather than being ancestral, or morphological differences resulting from adaptation to a different environment. By the late Miocene, the fossils recovered from Columbia, which had at least some tropical forest, begin to show clear links to the living species today.

HUMAN AND NONHUMAN-PRIMATE INTERACTIONS IN SOUTH AMERICA

The earliest definitive evidence of human presence in South America dates to approximately 12,500 B.P. at Monte Verde, Chile (Dillehay 1989, 1997; as previously noted, the earliest evidence for human presence in Amazonia is dated later, at approximately 11,000 B.P.). Assuming Amerindian arrival in the New World across the Bering Strait and the time needed for a southern migration across two continents, it is likely that the date for human presence will ultimately be pushed back to an earlier date. But in any event, there is roughly a 30-million-year gap between the arrival of the first primates in South America and the arrival of the first humans.

Little archaeological evidence is available regarding early human-nonhuman primate interactions in the Neotropics. However, Urbani and Gil (in press) have found interesting material from a cave site in eastern Venezuela dated to approximately 3000 B.P. Here, *Alouatta* remains are found in association with human tools, but the relationship is unclear. Cranial bones are found separate from and above other bones which suggests butchering. However, no traces of fire are found at the site. This suggests either that the monkeys were either cooked elsewhere (or perhaps eaten raw) or that they were associated with humans for purposes other than food.

Given that human- and nonhuman-primate sympatry extends more than 10,000 years in Amazonia, the abundance and geographic distribution of New World monkeys should take into account human influences. Several areas of human-nonhuman primate interactions are discussed below. These include direct redistribution of monkey populations by human beings, disease transmission, hunting pressure, habitat destruction, and use of anthropogenic forests.

Primate Trade

The geographic distribution of monkeys has been influenced by trade, in both pre-Columbian times and in more recent times. Although monkeys are naturally absent from coastal Peru, evidence of trade in monkeys can be seen as early as 3,000 B.C. from archaeological evidence of monkey effigies in stone, ceramics, and textiles, as well as in mummies in human burials (see Hershkovitz 1984:160–61). Further, based on biological evidence, Hershkovitz (1984) argued that the Central American *Saimiri oerstedi* demonstrates affinities to Amazonian forms and evidence of genetic drift due to inbreeding. He suggested that the species was introduced to Central America through pre-Columbian trade and that the same may be true for howlers, capuchins, and spider monkeys in Central America[1].

Old World Monkey populations are now established in several places in the New World due to human intervention. The vervet monkey (*Cercopithecus aethiops*) was brought from West Africa over three hundred years ago to the Caribbean Islands of St. Kitts, Nevis, and Barbados (McGuire 1974). In Florida, a group of Rhesus monkeys (*Macaca mulatta*) was released on an island on the Silver River for the benefit of a tourist cruise ride, but they escaped from the island and proliferated in the surrounding area (Wolfe 1997, 2002). In the Old World as well, primate populations have been directly redistributed by human beings. For example, the long-tailed macaque (*Macaca fascicularis*) population reached approximately nine hundred individuals on the island of Ngeaur in the Republic of Palau after the introduction of five pets during the German occupation around 1909 to 1914 (see Wheatley et al. 1997, 2002). Wheatley (1999:46–47) further suggests that the distribution of long-tailed macaques in Indonesia may be due to an ancient introduction of the species as pets.

In more recent times, the monkey trade in the Neotropics has tended to create pressure on populations rather than opening new niches. In addition to the export of monkeys as pets, they have been exported for medical research, zoo exhibits, and commercial byproducts. Neotropical primates were exported by Europeans since first contact with the New World. Written records date back as far as 1511 describing the transport of monkeys from the Río São Francisco in Brazil to Europe (see Urbani 1999). Hershkovitz (1972) reports that monkeys used as pets outside their country of origin are overwhelmingly New World monkeys, particularly *Saimiri* spp., *Aotus* spp., and *Cebus* spp. Soini (1972) points to 1950 as the date for the beginning of the systematic export of animals from Peru for use as pets and in biomedical research. By 1963, 30,000 monkeys were being exported per annum, with *Saimiri sciureus* representing almost 80 percent of those exported from Peru (Soini 1972:30).

Squirrel monkeys are the most commonly used Neotropical primate in medical research with projects ranging as widely as aerospace research,[2] Creutzfeldt-Jakob disease, infertility treatments, and research on Parkinson's disease (Abee 1989, 2000). In 1958 and 1959, squirrel monkeys were launched in Jupiter rockets to test the effects of space flight (Abee 2000). They have also been used to study the effects of diet and cigarette smoking (specifically carbon monoxide) on the development of coronary artery disease (Abee 1989, Clarkson and Klumpp 1990). Squirrel monkeys have been used in research involving human parasitic diseases, including *Trypanosoma cruzi*, which occurs naturally in squirrel monkeys, and other parasitic diseases with which they can be experimentally infected, including toxoplasmosis (*Encephalitozoan cuniculi*), the blood parasite *Babesia microti*, and *Leshmania donovani* (the squirrel monkey is a nonhuman host). They can serve as hosts for eight known schistosomes (Galland 2000). More recent studies have included malarial vaccine development, experimental infection with Creutzfeldt-Jacob disease, emphysema, and reproductive studies of hormone-induced ovulation and artificial insemination (Abee 2000). Squirrel monkeys are particularly important in malarial research because they can be infected with all four of the malarial protozoa that afflict human beings: *Plasmodium vivax*, *P. falciparum*, *P. ovale*, and *P. malariae* (Galland 2000, Sibal and Samson 2001).

Most squirrel monkeys used for research in the 1960s and 1970s came from Brazil and Colombia, until those countries banned their export (Abee 1989). In at least one area of Colombia (Chiriquí), *Saimiri sciureus* populations were reported to be depleted due to hunting for export (Baldwin and Baldwin 1976). Researchers then began to rely on Peruvian and Bolivian squirrel monkeys (*S. boliviensis*); after Bolivia banned export in the 1980s, researchers used *S. sciureus* from Guyana until 1989, when they stopped exporting the monkeys (Abee 1989). By the late 1990s, only a small number continued to be exported from Peru, with Surinam providing additional exports (Abee 2000). The endangered

Costa Rican squirrel monkey, *Saimiri oerstedi oerstedi*, was also exported for medical research in the past (Rylands, Mittermeier, and Luna 1997).

Owl monkeys (*Aotus* spp.) and capuchin monkeys (*Cebus* spp.) have also been exported for use in medical research. Owl monkeys have been popular for ophthalmic (Ogden 1994) and malarial research (Collins 1994, Sibal and Samson 2001). Between 1968 and 1972, the United States imported over 20,000 owl monkeys, primarily from Colombia and Bolivia, but approximately two-thirds had died before being exported, so the number actually captured was closer to 60,000 (see Aquino and Encarnación 1994:90). In Colombia, both *Cebus albifrons* (Green 1976) as well as *C. capucinus* (Baldwin and Baldwin 1976) have been captured for export with populations of the latter reported as having been depleted in Chiriquí, Colombia. In addition to their use in general biomedical research, more recently *Cebus* monkeys are being trained by the Boston based "Helping Hands" program to perform manual tasks and to be companions for quadriplegics.

The woolly monkeys (*Lagothrix lagotricha*) have been a particularly popular pet species within South America and have also been heavily exploited for export as pets (Mittermeier 1987b). Alfred Wallace kept several as pets during his South American travels (Fooden 1963). They have commanded the highest prices in the pet trade in Peru (Soini 1972). In several regions of Colombia, it has been reported that adults are killed for food while the infants may be sold as pets (Bernstein 1976; Green 1976; Hernández-Camacho and Cooper 1976). At least 2,500 woolly monkey were sold annually in Iquitos, Peru, until the 1973 ban (personal communication from Neville cited in Heltne and Thorington 1976:122). Currently, they are critically endangered (Rylands, Mittermeier, and Luna 1997).

Other cebid monkeys also have been utilized to a lesser extent. A variety of spider monkey species have been reportedly sold as pets in Colombia and Peru (Bernstein 1976; Green 1976; Neville 1976; Soini 1972). Near the Amazon river in Southern Peru, spider monkeys (*Ateles paniscus*) have been hunted for export (Durham 1975). The vulnerable *Pithecia monachus* has been exported in small numbers from Leticia, Colombia (Hernández-Camacho and Cooper 1976; Rylands, Mittermeier, and Luna 1997), and howlers (*Alouatta seniculus*) also have been exported in Colombia (Green 1976).

Among reports of callitrichid trade, the common marmoset (*Callithrix jacchus*) has been one of the most commonly utilized New World monkey for medical research (Bowden and Smith 1992). *Saguinus oedipus* were hunted heavily for export in Colombia until a 1969 law was enacted to protect them (Hernández-Camacho and Cooper 1976; Neyman 1978); however, they remain endangered (Rylands, Mittermeier, and Luna 1997). There are reports that

Saguinus leucopus has also been traded for export in northern Colombia (Green 1976), and they are still considered vulnerable (Rylands, Coimbra-Filho, and Mittermeier 1993). In Peru, over 22,000 *Saguinus fuscicollis* were exported between 1962 and 1973, and they were also traded domestically as pets (Freese, Freese, and Castro 1978:128). According to Mittermeier, Bailey, and Coimbra-Filho (1978), *Saguinus* spp. have often been brought from the Brazilian Amazon across the border to animal dealers in Leticia, Colombia. It has also been reported that *Saguinus* spp. have been captured in Colombia and Peru for export for biomedical research (Hernández-Camacho and Cooper 1976; Freese, Freese, and Castro 1978). *Cebuella pygmaea* have been hunted for export as pets in Colombia and Peru, but they have been protected by law in Peru since 1970 (Freese, Freese, and Castro 1978; Moynihan 1976b; Ramirez, Freese, and Revilla 1978). Recently, the critically endangered golden-headed lion tamarin (*Leontopithecus chrysomelas*) from the Mata Atlântica (the Atlantic rain forest) in Brazil has been included in the illegal Brazilian and international trade (Mittermeier 1987c; Rylands, Mittermeier, and Luna 1997).

Commercially, and for the most part on a local basis, monkey products have been sold for a variety of non-dietary purposes. Stuffed monkeys and ornaments of monkey bones are sold to tourists in Brazil (Mittermeier and Coimbra-Filho 1977). The hyoid bone of *Alouatta seniculus* is used by campesinos in the upper Magdalena and Cauca valleys of Colombia as a grinding device, and some believe that it possesses therapeutic properties for curing goiters (Hernández-Camacho and Cooper 1976). In some parts of the Brazilian Amazon, drinking from an *Alouatta* hyoid has been thought to ease labor pains, and in Surinam, it has been thought to cure stuttering (Mittermeier and Coimbra-Filho 1977)[3]. There have been reports that *Alouatta seniculus* skins have been sold as well (Hernández-Camacho and Cooper 1976). Near the Amazon river in Southern Peru, *Ateles paniscus* has been hunted for its pelt (Durham 1975), and in Colombia, the fur of *Aotus* spp. has been used to decorate bridal wear (Hernández Camancho and Cooper 1976). *Chiropotes* spp. and *Pithecia* spp. have been killed for their thick tails, which are used as feather dusters in Brazil and Surinam (Husson 1957, cited in Wolfheim 1983; Mittermeier and Coimbra-Filho 1977; Soini 1972). While Neotropical primates and their products continue to be exported and traded locally for various purposes, their export seems to have declined significantly over the last decades. Bowden and Smith (1992) estimate that there has been a 90 percent decrease in the number of monkeys exported between the 1960s and the 1990s.

It should also be noted that trade has also affected Old World primate populations. For example, colobus monkeys began being heavily exploited in the latter half of the nineteenth century by the international fur trade, leading to the extinction of some populations (Oates 1977). Mother macaques in Ngeaur, Republic of Paulau, are sometimes killed so their infants may be sold as pets (Wheatley et al. 1997), although recent legislation imposing a hefty fine for the illegal export of monkeys seems to be an effective deterrent (Wheatley et. al 2002). The capturing and smuggling of infants are problems for the Eastern lowland gorilla as well (Goodall and Groves 1977). In addition, there is currently much concern over the African "bushmeat crisis," which involves the illegal commercial trade of African apes (see Hutchins 1999; Rose 2002; also see chapter eight).

Zoonoses and Anthroponoses

Another dimension of human-nonhuman primate interactions is the close genetic relationships among primate species, which have allowed some infectious organisms to jump from their natural hosts to other primate species. Perhaps the most notorious pathogen involves the long-suspected simian origin of human HIV (Myers, MacInnes, and Myers 1993). It has been confirmed only recently that the virus originated in the common chimpanzee, *Pan troglodytes troglodytes* (Gao et al. 1999). Also, one laboratory worker allegedly contracted SIV (simian immunodeficiency virus) from an experimentally infected macaque (Khabbaz et al. 1994). A non-pathogenic AIDS-like virus has also been found in Brazilian Amazonian Indians that may have been transmitted from Neotropical monkeys to humans (see Feldman 1990, quoting Francis Black). In 1989, a filovirus was discovered in some long-tailed macaques exported from the Philippines that was initially feared to have the potential to spread to human beings (Bowden and Smith 1992). Ebola filoviruses can infect both monkeys and human beings, and ebola is suspected to be a simian zoonosis, although this has not yet been proven (Peters et al. 1993). A Center for Disease Control and National Institutes of Health voluntary testing program also found that 1.3 percent of workers in primate biomedical research had been exposed to nonhuman primate spumavirus ("foamy virus") (Center for Disease Control 1997).

Cross-species infections occur between New World monkey species, particularly when housed together for research purposes. But a similar environment is created when indigenous peoples keep several monkey species in close proximity as pets. Further, human beings such as the Guajá, living in close proximity to monkey pets, provide potential opportunity for zoonotic infections in humans or, conversely, anthroponotic infections in monkeys.

Several examples of primate-to-primate disease transmission have come from the herpes viruses. *Herpesvirus ateles* often infects spider monkeys asymptomatically, but it causes lymphoma in some species of owl monkeys and tamarins (King 1994). *Herpesvirus saimiri* (*H. tamarinus*) is carried asymptomatically in squirrel and spider monkeys but is typically lethal in two to six days in owl monkeys and can cause lymphoma when experimentally induced in tamarins, marmosets, capuchins, and spider monkeys (Weller 1994; King 1994). Owl monkeys in captivity can contract the herpes simplex virus (*Herpesvirus hominis*) from humans, which produces a severe disease in them (Weller 1994:186). Old World monkeys and apes carry herpes viruses that can be passed to human beings. When the virus enters a closely related host, the pathogenesis of the disease can be much more severe than what occurs in the natural host, and it can even take a lethal course (Eberle and Hilliard 1995). The herpes B virus (*Herpesvirus simiae*), is common in macaques but leads to lethal encephalitis in approximately 78 percent of those developing symptoms (CDC 1987a, 1987b; Fiennes 1967). The risk of symptomatic infection appears to be extremely low, with less than thirty documented symptomatic cases of transmission of herpes B from rhesus macaques to humans (Wolfe 2002) despite at least 70 percent of macaques' harboring the virus (CDC 1998) and thousands of human beings' having had direct contact with them (CDC 1987b). Most cases have occurred after a macaque bit or scratched an animal worker in a laboratory research setting. However, in 1987, a case of human-to-human transmission was documented, from a husband working with macaques in a Florida medical research facility to his wife (CDC 1987a). It is reported that the wife had a break in the skin integrity of her hands due to dermatitis and that she had been applying topical medications to her husband's herpetic lesions. In addition, in 1998, a twenty-two year old worker in a Georgia research facility developed fatal encephalitis subsequent to macaque body fluids splashing her in the eye (Center for Disease Control 1998). This was the first known case of mucocutaneous transmission to humans in the absence of an injury or other break in the skin integrity. In addition, other potential zoonoses include the less-well-known simian alpha-herpesviruses (Eberle and Hilliard 1995).

Although simple sympatry of human and nonhuman primate species may pose a low risk for zoonoses or anthroponoses, indigenous pet-keeping may pose risks even greater than those experienced in a research environment. The Guajá premasticate foods for infant monkeys and allow them to eat directly from their mouths, providing a direct route for oral–herpes virus transmission and potentially for other diseases. During the research period, transmission of what appeared to be oral herpes (*Herpes simplex*) was observed among two young boys who were parallel cousins and their pet monkeys, a Ka'apor capuchin and a howler. In both boys, large fluid-filled blisters developed on and around their lips, which scabbed and resolved within a two-week period. The capuchin also developed such lesions on and around her lips, while the howler developed the lesions on her entire face. Both the capuchin and the howler died following the infection, suggesting anthroponotic infection in the monkeys. Anecdotally, the Guajá also report that their monkeys die from the same respiratory infections they themselves contract[4].

Their report may very well be accurate given that respiratory infections accounted for 24 to 27 percent of all owl monkey deaths over an eight year period at the Center for Reproduction and Conservation of Nonhuman Primates in Iquitos, Peru (see Weller 1994:198). Infections with influenza-A viruses have been induced experimentally in squirrel monkeys, which created symptoms similar to those seen in humans (B. Murphy 1993). Further, tuberculosis and measles are well-known anthroponotic infections that can occur when monkeys and apes come in contact with humans, particularly in zoos, in research environments, and when nonhuman primates are kept as pets (Renquist and Whitney 1987). Other diseases afflicting humans that have been documented in nonhuman primates are scabies (chimpanzees, gorillas), yaws (baboons), schistosomiasis (baboons), pneumonococcal pneumonia (great apes), and polio (orangutans, chimpanzees) (Wallis and Lee 1999).

Human-nonhuman primate interactions may also play a role in the transmission of yellow fever and malaria. Yellow fever is believed to have been introduced to the New World from Africa through the slave trade (K. M. Johnson 1993; Morse 1993). Periodic yellow fever epidemics were documented in howlers on Barro Colorado Island until deforestation isolated the monkeys (Moynihan 1976a). Evidence exists for a close relationship between the two monkey malarias, *Plasmodium brasilianum* and *P. simium*, and two human malarias, *P. malariae* and *P. vivax*, respectively (Collins 1994). Stewart et al. (1998) state that both species of *Plasmodium* should be considered as actual or potential zoonotic infections. I would further add that the human malarias should be considered potential anthroponotic infections for monkeys as well. In addition, transmission of intestinal parasites between howlers and human beings has been described by Stewart et al. (1998), including human *Giardia intestinalis* infecting howlers and howler tapeworms infecting human beings; howler roundworms are also a potential zoonotic infection.

Hunting Pressure

Sponsel (1997:152) has argued that human predation on nonhuman primates has been "grossly neglected," despite evidence for a predatory relationship since the time of the australopithecines. A number of archaeological sites have yielded evidence of hominid association with nonhuman primates, with primates at times the dominant fauna (e.g., Arambourg 1955; Cook 1963; Howell, Fichter, and Eck 1969; Leakey 1965; Woo 1966). However, an association does not necessarily indicate hunting, as Binford (1977) argued regarding the Acheulian hand axe site at Olorgesailie, Kenya, dated by Isaac (1977) at 400,000–700,000 B.P. However, a recent study of the bone breakage patterns at the Kenyan site provides evidence of systematic butchery by *Homo erectus* of the giant gelada baboons (*Theropithecus oswaldi*), which were the dominant fauna in the area (Shipman, Bosler, and Davis 1981). Further, Groves (1981) has suggested that the giant gelada baboons and *Homo erectus* may have had similar diets and that the baboons may have been killed as competitors and consequently eaten as food.

Among some chimpanzees (*Pan troglodytes*), evidence exists that their primate predation can affect the group size and distribution of other primate species. Recent attention has been given to primate-primate hunting in Africa, where primate hunting by chimpanzees is now understood to be basic to their behavioral ecology. At Gombe, Stanford (1995, 1996, 1999) found that red colobus monkeys (*Colobus badius*) were the major prey animal of chimpanzees, accounting for 50 to 80 percent of vertebrate species killed for food. Further, within the core hunting area of the chimpanzees, red colobus group size is 46 percent smaller than at the periphery. Further, Wrangham and Bergmann Riss (1990) have suggested that as much as 41 percent of the red colobus monkey population is eaten yearly by chimpanzees at Gombe.

Human beings may have been responsible for the extinction of many species of primates. In the Old World, human beings have been implicated in the extinction of most of the large monkeys from East Africa during the early and middle Pleistocene, the disappearance of the orangutan from mainland Asia, and the extinction of large diurnal lemurs in Madagascar (see Fleagle 1988). In the New World, human beings may have also been responsible for the extinction of some species. *Xenothrix mcgregori*, for example, is suspected to have been exterminated by the earliest human colonizers of the Antilles (Hershkovitz 1972). Further, the disappearance of the previously described *Protopithecus brasiliensis* roughly coincides with human occupation in the region, which may indicate human influence, following P. S. Martin's (1984) model of New World extinctions in the terminal Pleistocene. However, climatic and subsequent environmental changes may have ultimately caused the decline of this species.

Evidence exists for human predation of virtually all monkey species by both indigenous populations and more recent settlers in the Neotropics. Peres (1999) estimates that three to eight million primates are consumed each year by the rural populations in the nine states of the Brazilian Amazon. The larger species tend to be targeted more often, particularly howlers, spider monkeys, woolly monkeys, and capuchins (see Mittermeier 1987a). Some monkeys may be more threatened by hunting than by habitat destruction. The species described include *Alouatta seniculus* and *Cebus albifrons* in Trinidad (Neville 1976), *Alouatta palliata* and *Ateles geoffroyi* in Mexico (Leopold 1959), *Ateles paniscus* in Colombia (Hernández-Camacho and Cooper 1976), and monkeys in general in the Peruvian Amazon near human settlements with populations of one thousand or more (Aquino and Encarnación 1994). Medium-sized monkeys may be hunted more extensively once the larger species have been hunted out, as is the case with *Saimiri sciureus* in Peru (Soini 1972). While the callitrichids are hunted, they tend to be much less intensively hunted (e.g., Freese, Freese, and Castro 1978; Mittermeier, Bailey, and Coimbra-Filho 1978). Further, Eisenberg (1978) pointed out that because of twinning and faster reproductive and maturation rates, callitrichids are better able to withstand and recover from human exploitation than other New World monkeys. It should also be mentioned that monkeys are sometimes hunted and used as bait to attract large felines (see Wolfheim 1983).

In order to evaluate hunting pressure on monkey populations, several important considerations should be taken into account. First, the type of technology used plays a role in the degree of hunting pressure on monkeys. For example, among the Matsigenka of Peru, areas with shotgun technology have experienced local extinctions of the larger monkeys (woolly and spider species), but the larger monkeys remain plentiful in areas where only indigenous technology is used (see Shepard 1997, 2002). In the Old World, Wheatley (1999) found that shotgun hunting in the Republic of Ngeaur affected long-tailed macaques (*Macaca fascicularis*) by increasing the number of troops, decreasing the number of individuals within a troop, and reducing the overall population by half. He makes the important point here that these results demonstrate that human behaviors can affect nonhuman primate social organization.

In understanding the effects of hunting pressure, it is also important to consider that often hunting pressure (particularly with indigenous technologies) in itself is not the real threat to monkey populations, but may become a serious problem in the wake of deforestation. When monkey populations are reduced and restricted to circumscribed patches, they can be quickly hunted out entirely. For example, hunting pressure due to habitat destruction has been described for the Peruvian yellow-tailed woolly monkey (*Lagothrix flavicauda*) in the montane forest of Peru (Mittermeier and Coimbra-Filho 1977). Lizarralde (1997, 2002) also asserts that the disappearance of spider monkeys from part of the territory of the Barí of Venezuela is primarily due to the circumscription of their habitat by deforestation, which then allowed them to be quickly hunted out. In sum, the problem may not be so much that hunting itself is unsustainable, but that deforestation makes hunting unsustainable, for it affects not only nonhuman primate communities, but also threatens societies that traditionally use primates as food.

Human Influence on Nonhuman-Primate Habitats

Numerous researchers have described threats to nonhuman primate populations due to deforestation in both the Old and New Worlds.[5] According to Mittermeier, Kinzey, and Mast (1989), habitat destruction is the most significant influence on primate populations. They describe agricultural development of the tropical forests as a major problem, not only because 90 percent of primate species live in the tropical forests, but because poor soils make agriculture difficult to sustain for more than a few seasons, which has led to ongoing development of the forests.

Several New World species provide evidence of threats from deforestation; a sampling includes *Aotus infulatus* (Aquino and Encarnación 1994), lion tamarins (*Leontopithecus rosalia*) and *Alouatta fusca* in southeastern Brazil (Chiarella 1993; Coimbra-Filho and Mittermeier 1977), *Cebus apella* in the Colombian llanos (Mittermeier and Coimbra-Filho 1977), *Cebus capuchinus* and *Saimiri sciureus* in Panama (Baldwin and Baldwin 1976), *Alouatta pigra* and *Alouatta palliata* (Crockett and Eisenberg 1986), and *Brachyteles arachnoides* (Kinzey 1982). Old World monkeys and apes are also threatened, including *Colobus guereza* (Oates 1977), *Presbytis entellus* in India (Oppenheimer 1977), gibbons and siamangs of southeast Asia (*Hylobates* spp. and *Symphalangus*) (Chivers 1977), and Eastern gorillas in the Republic of Congo (formerly Zaire) (Goodall and Groves 1977). Although deforestation is usually due to development or agricultural use, in Vietnam the major source of forest degradation threatening Douc langurs (*Pygathrix nemaeus*) are the byproducts of extended warfare, including plant damage from defoliants, such as Agent Orange, and landscape scarring from munitions and millions of bomb craters (see Lippold 1977:515).

Habitat alteration can have a direct effect on the social behavior of primate species. Long-term studies with *Alouatta palliata* at La Pacifica suggest a relationship between environmental change and intergroup relatedness (e.g., Clarke, Glander, and Zucker 1998; Clarke and Zucker 1994; Clarke, Zucker, and Glander 1994; Clarke and Glander 1984; Zucker and Clarke 1998; also see

Eisenberg, Muchkenhirn, and Rudran 1972). The studies describe *Alouatta palliata* with both male and female emigration patterns and with male infanticide and female aggression towards non-related kin. More emigration occurs in unstable environments, and this possibly affects intergroup relatedness and, thus, the level of agonistic behaviors towards non-related infants. As Clarke, Glander, and Zucker (1998) point out, group size and composition here is not species-specific but must be understood as responsive to the environmental change. The implication here is that habitat destruction by humans can affect nonhuman primate phenotypes. A somewhat similar example comes from the chimpanzees of Gabon. It has been reported that logging triggered a "chimpanzee war" by causing competition over territory in the reduced habitat (Stevens 1997).

While the devastating effects of humans' deforesting the habitats of nonhuman primates have been addressed by numerous researchers, little attention has been given to other forms of habitat modification affecting human-nonhuman primate ecological relations. Sponsel (1997), however, has suggested that swidden agriculture may be important to some monkey species that are either attracted to swidden gardens or to the secondary growth in old fallows. With Balée's (1989) estimate that at least 12 percent of the Amazon is anthropogenic forest, nonhuman primate adaptation (as well as all faunal adaptation) to anthropogenic forests merits further attention.

In the most extreme anthropogenic environment, there is total elimination of the natural habitat of nonhuman primates. In some cases, however, human interactions with nonhuman primates through food provisioning have preserved monkey populations that would not have survived otherwise in the wake of deforestation. In Jaipur, India, development has caused the forest to disappear; Rhesus monkeys are able to survive because they are provisioned by the local people for whom the monkeys figure in their religious beliefs (Wolfe 1991, 1992). Wolfe argues further that in India as a whole, as well as in Japan and possibly in China, monkey populations would not survive without human provisioning. In addition, Wheatley (1999) argues that the "natural" habitat of some Old World primate species may involve close association with human beings. In one study he cites, long term observation of Rhesus macaques in north-central India demonstrated that 86–88 percent of the population lives in commensal or semi-commensal habitats with human beings (Wheatley 1999:39–40).

Agricultural gardens and fields are another type of anthropogenic environment that has affected nonhuman primate species. Domesticated plants provide a new food resource for some primates which may influence their geographic range and population size. The most notorious primate crop pests in the New World are the capuchins, which are often killed to keep them away from gardens. Crop-raiding has been reported from *Cebus apella* and *Cebus albifrons* in

Colombia (Green 1976; Klein and Klein 1976; Moynihan 1976a; Hernández-Camacho and Cooper 1976), *Cebus capuchinus* in Panama (Baldwin and Baldwin 1976), and *Cebus* spp. monkeys in Maranhão, Brazil (Balée 1994a). *Cebus apella* is even called *maicero* (maize eater) in Colombia and several other parts of South America (Moynihan 1976a:106). *Saimiri sciureus* has been reported as an occasional crop-raider in Chiriquí, Panama, and is killed as a crop pest (Baldwin and Baldwin 1976). Some callitrichids also have been reported as crop pests, including *Callimico goeldii* (Moynihan 1976a), *Callithrix jacchus* (Hershkovitz 1977), *Saguinus leucopus* (Green 1976), *Saguinus mystax*, and *S. fuscicollis* (Castro and Soini 1978).

Similar relationships have been reported in the Old World. Long-tailed macaques (*Macaca fascicularis*) raid gardens in Bali (Wheatley 1999), and in the Republic of Ngeaur, they are killed as agricultural pests because they raid and damage gardens and even raid houses for food (Wheatley et al. 1997, 2002). Further, black colobus monkeys (*Colobus guereza*) also have been killed as crop pests in Africa (Oates 1977). The Barbary macaque (*Macaca sylvanus*) raids crops and human garbage in Morocco and is killed to prevent their stripping cedar bark, which damages *Cedrus libanotica atlantica* trees (see Deag 1977). Rhesus macaques (*Macaca mulatta*) are crop raiders in Northern India (Southwick and Siddiqi 1977). Baboons (*Papio* spp. and *Therapithecus gelada*) are crop raiders, and some gelada baboons glean abandoned threshing floors in Ethiopia (Dunbar 1977; Forthman-Quick 1984; Strum 1984). One study found that baboons were responsible for 70 percent of crop damage by animals among farmers in the Budongo National Forest in Uganda (Hill 2000).

Just as hunting pressure on monkeys may become a serious problem only with deforestation, Oppenheimer (1977) has noted that monkey encroachment into human habitats for crop raiding is often the result of loss of habitat. Dunbar (1977) also has reported an association between gelada baboon crop raiding and habitat destruction. Among the Toque macaques (*Macaca sinica*) in Sri Lanka, in areas where natural habitat had been cleared for bovine pasture, monkeys raided gardens and human garbage, but those in less disturbed habitats tended to avoid human areas even when their habitat bordered cultivated areas (Dittus 1977). Similarly, in Japan, 6000 hectares were damaged by crop-raiding macaques, and, in 1998 alone, over 10,000 macaques (*Macaca fuscata*) were culled as nuisance pests (Sprague 2002). Sprague suggests that the increase in macaque crop raiding may actually be the result of changes in land use related to the decline in farming as a lifestyle. He suggests that the absence of human activities around the farm have eliminated a buffer zone between the forest and the field that could act to deter monkeys from approaching the farms.

Old fallow forest or the succeeding secondary forests provide another anthropogenic environment that nonhuman primates can exploit. Balée (1994a) has reported that *Cebus* monkeys forage in recently abandoned Ka'apor Indian settlements for fruits. In Colombia, *Cebus capucinus*, *C. albifrons*, and *C. apella* are able to live in secondary forest. *Cebus* spp. in general are adaptable and able to exploit old, fallow, secondary forest, and a variety of other habitats (Izawa and Bejarano 1981; Mittermeier and Coimbra-Filho 1977). However, *Cebus albifrons* prefers undamaged high forest (Izawa and Bejarano 1981).

In addition, *Cebus apella* heavily exploits palms, some of which grow in old fallows. For example, Terborgh (1983) has observed *Cebus apella* exploiting *Astrocaryum* palms, which grow in old fallows and are traditionally important fibers for Guajá clothing and hammocks. Further, the Guajá report that *Cebus apella* is able to exploit the babassu palm (*Orbignya phalerata*), which is a traditional staple of the Guajá (also see Balée 1988). They state that the monkeys pound them open. Pet *Cebus apella* monkeys among the Guajá were observed on numerous occasions pounding babassu fruits but were never observed to open one successfully. These palm nuts are extremely hard, but it is possible that the monkeys are capable of opening them due to their thick dental enamel (thicker than any living primate) and suite of craniomandibular features that allow high degrees of masticatory stress (Cole 1992, Janson and Boinski 1992). It is also possible that the capuchins exploit older nuts. Agoutis (*Dasyprocta* spp.) and pacas (*Agouti paca*) eat the mesocarp (May et al. 1985), and such partially eaten babassu fruits were commonly seen in the Guajá areas. A combination of pre-processing by rodents and exposure to the elements may soften the endocarp of the fruits making it easier for the capuchins to reach the kernel. In addition, bruchid beetles (*Pachymerus nucleorum*) enter fruits on the ground and feed on the kernels (May 1985 et al.), and the capuchins may actually be exploiting the palm grubs growing in the older nuts.

Several other monkey species are able to exploit secondary forest. In Colombia, Bolivia, as well as in the Brazilian Amazon, *Aotus* spp. have been observed in secondary old fallow forest and near human habitations (Izawa and Bejarano 1981; Hernández-Camacho and Cooper 1976; Mittermeier and Coimbra-Filho 1977). *Saimiri sciureus* can exploit old fallow and *terra firme* secondary forest and are frequently seen close to human habitations (Abee 1989; Izawa and Bejarano 1981; Mittermeier and Coimbra-Filho 1977). *Alouatta* spp., when not under extreme hunting pressure, can survive in forest fragments, disturbed habitats, old fallow, and in close proximity to human populations, although they may suffer adverse health effects if their range is limited to a small fragment (Crockett 1998; Horwich 1998; Izawa and Bejarano 1981; Mittermeier and

Coimbra-Filho 1977). *Alouatta fusca* and *Alouatta seniculus* have been observed to range even into urban areas (see Crockett 1998).

The callitrichids are often found in the secondary forest, and some species even prefer it. In Colombia, *Saguinus nigrocollis, S. fuscicollis, S. oedipus* are often found in the secondary forest, and *S. geoffroyi* and *S. leucopus* are reported to prefer secondary forest (Hernández-Camacho and Cooper 1976). Mittermeier, Bailey, and Coimbra-Filho (1978) found similar results in their survey of the Brazilian Amazon, where callitrichids were commonly found in the *terra firme* secondary forest. *Saguinus midas* and *S. mystax* preferred secondary growth near human habitations. They suggest that areas near human settlements are beneficial to the callitrichids because predatory birds are often scared away, and larger primate competitors are often hunted out locally. *Saguinus bicolor* has been observed within the city limits of Manaus in the secondary forest, and *Saguinus midas niger* (the subspecies kept by the Guajá) has been observed in close proximity to Belém, the largest city in the Amazon region (Mittermeier, Bailey, and Coimbra-Filho 1978). *Callithrix* spp. in general survive well in close proximity to human settlements and in the secondary forest (Izawa and Bejarano 1981; Mittermeier and Coimbra-Filho 1977). *Callicebus* spp. also are able to live near human settlements (Mittermeier and Coimbra-Filho 1977).

A few species can be characterized as thriving in anthropogenic habitats. In the Old World, Wheatley (1999) found that long-tailed macaques (*Macaca fascicularis*) in Kalimantan, Indonesia, used secondary forests, largely created by swidden agriculture, as the core area of their home range (Wheatley 1999:42). Richard, Goldstein, and Dewar (1989: 573) distinguish between "weed macaques" and "non-weed macaques," with the former described as thriving in habitats disturbed by humans.

In the New World, several researchers have argued that *Saguinus geoffroyi* may have moved into forests of the Atlantic coast of Panama only in the last few thousand years in the wake of pre-Columbian swidden agriculture (Bennett 1968; Moynihan 1976a, 1976b). Here, old fallows promoted the growth of secondary forest that *S. geoffroyi* exploited by moving from patch to patch into eastern Panama. In Colombia, *Cebuella pygmaea* has been reported to be well adapted to hedgerow clumps of degraded woods between pastures and crop fields (Moynihan 1976a, 1976b). Moynihan (1976b) has gone so far as to call *Cebuella pygmaea* a commensal of human beings, also suggesting that the species benefits from predator protection by being near human settlements.

Several species, however, seem unable to exploit anthropogenic habitats. *Lagothrix* spp. (Bernstein et al. 1976; Hernández-Camacho and Cooper 1976; Izawa and Bejarano 1981), *Cacajao* spp. (Mittermeier and Coimbra-Filho 1977), and *Brachyteles arachnoides* (Kinzey 1982) do not survive well in the secondary or remnant forests. *Ateles* spp. also prefer *terra firme* primary forest (Izawa and Bejarano 1981; Mittermeier and Coimbra-Filho 1977). *Chiropotes* spp. had been thought to have difficulty adapting to secondary forest (Mittermeier and Coimbra-Filho 1977), but one recent study observed *Chiropotes satanas utahicki* occupying fragmented forest more commonly than the primary forest (Bobadilla and Ferrari 2000). This suggests they may have a higher tolerance for habitat disturbance than previous studies have suggested. Many species of these genera are endangered. Thus, descriptions of primate sympatry should involve human beings, since some nonhuman primates are sympatric with some human cultural strategies of forest usage, even though others are allopatric and cannot exploit the same habitat as humans.

Another area of potential ecological relationship between human and nonhuman primates involves seed dispersal. Balée (1994a) has reported that *Cebus* monkeys forage in recently abandoned Ka'apor Indian settlements for fruits and act to disperse species in the surrounding swiddens that will develop into old fallow. Further, according to Balée, the Ka'apor report *Cebus* monkeys as specific dispersers of seeds from wild cacao and ingá trees. Terborgh (1983) also reported *C. apella* and *C. albifrons* exploiting *Theobroma cacao*. Howlers in the Atlantic forest are seed dispersers of several plant species edible by human beings, including the hog plum (*Spondias mombin*), *Annona* sp., and *Campomanesia* sp., (Bonvicino 1989; see Cavalcante 1996 and Balée 1994a). Bonvicino (1989) also reported howlers as dispersers of *Protium heptaphyllum*, which is both an important old fallow species (Balée 1994a) and is eaten by the Guajá. Julliot (1996) found that red howlers (*Alouatta seniculus*) in French Guiana dispersed more than 95 percent of the seeds of the ripe fruit they ate and that passage through the digestive tract did not significantly modify the germination success of most plant species. *Chiropotes satanas* may also act as a seed disperser, since it eats the mesocarps of *Spondias mombin* and *Protium tenuifolium*, both of which people also consume (Norconk, Wertis, and Kinzey 1997; see Cavalcante 1996 and Balée 1994a). *Cebus apella* and *Saimiri sciureus* also eat fruit of the hog-plum (*Spondias mombin* L.) (Terborgh 1983).

CONCLUSION

Historical ecology has the potential to offer insights into the ecological relationships of humans and nonhuman primates in the Neotropics. In the following chapter, the ecological relationships between monkeys and the Guajá are discussed. The Guajá forage in anthropogenic forests for their caloric staple, the babassu palm. These anthropogenic forests are also an environment that is attractive for the monkeys they hunt. *Cebus apella* heavily utilizes palms as well as other secondary forest species and can be considered, like the Guajá, to be attracted to old fallows and to act as a dispersal agent for some plants edible to humans. *Alouatta* spp. exploit secondary forest, utilize old fallow colonizers for food, and act as seed dispersers for plants edible by humans. *Saimiri sciureus* is able to exploit some old fallow plants, and both the squirrel monkey and *Aotus* spp. are able to exploit secondary forest. *Saguinus midas* is able to live near human settlements and may even prefer this habitat due in part to the protection it offers from nonhuman predators, such as felines, birds of prey, and other carnivores.

The Guajá indigenous reserves also provide a sanctuary for the endangered black-bearded saki (*Chiropotes satanas satanas*) and the newly identified Ka'apor capuchin (*Cebus kaapori*). As mentioned above, *Chiropotes* spp. may not survive well in secondary forest, so preservation of the indigenous reserves provides a refuge for the species. The same may be true for *Cebus kaapori*, whose habitat is circumscribed to the vicinity of the Guajá indigenous reserves.[6] Although the Guajá hunt this species, that hunting seems to be sustainable since the reserves provide areas where the species continue to exist.

Human trade, hunting patterns, and habitat utilization can affect dramatically nonhuman primate ecology and behavior. The review of the literature here suggests this holds true wherever human and nonhuman primates coexist. Clearly, human ecology and nonhuman primate ecology are intertwined. Awareness of these relationships is important in understanding local ecological systems, and, perhaps more importantly, issues of primate conservation should address the cultural practices and perceptions of the people involved.

3

MONKEY HUNTING

Broadly, monkeys can be considered a widely available food source in lowland South America.[1] Where quantitative studies have been done, the degree to which monkeys contribute to a given group's diet varies from less than 3 percent to 36 percent (e.g., see Berlin and Berlin 1983; Kaplan and Kopischke 1992). Numerous interrelated ecological factors may contribute to these differences, including the specific types of species available, species densities in a given area, population density of a given group, means of extraction, local habitat differences, and historical changes in the environment and hunting technologies of a group.

However, ecological determinism is insufficient to account for all of the differences. Local perceptions also come into play, such as the disparate food taboos that appear among groups. For example, monkeys are taboo for pregnant Yanomamö women and their husbands, adolescent Jivaro girls for twelve months after their first menses, preadolescent Desana boys, and all Kayapó women (McDonald 1977). Shepard (2002) describes howlers as the most abundant mammal in Manu Park, Peru. However, Matsigenka hunters take howlers ten times less frequently than the similarly sized woolly and spider monkeys in the region. Thus, the material importance of monkeys to a given lowland South American culture cannot be presumed to be determined simply by the environmental availability of monkeys, but must be investigated by understanding the culture as a whole. As Rival (1998) argues, the food quest is embedded in sociopolitical processes.

FIGURE 3.1 Hunting howlers (Alouatta belzebul belzebul)

As previously mentioned, early contact reports with the Guajá suggested that the babassu palm was their key plant staple and that howlers were a key game animal. Meirelles's (1985) description of a recently contacted group described howlers as being integral to the diet, an *"alimentação fundamental."* An account of the first contact with a group of Guajá in the Araribóia area described camps littered with many monkey skeletal remains, in addition to tortoise carapaces (Carvalho 1992). In the research conducted here, three major patterns of animal exploitation were found: howlers as the key animals food source in the wet season, fish as the key animal food source in the dry season, and white-lipped peccary hunting playing an important role in both the wet and the dry seasons (also see Cormier 2002b).

FOOD TABOOS

The Guajá have three types of food taboos: foods that no one may eat, foods that interfere with fertility, and foods that children avoid (see table 3.1). The Guajá describe foods that may not be eaten as *i'upɔtawa* and those that may be eaten as *i'upɔrɔ*. The primary reason given for not eating a particular food is that it is *ka'akačɔ'ɔ* (strong smelling) or *inamahɔ* (bad smelling). In some cases, the Guajá simply said that they were *kiye* (afraid of the food) or *nɔkwai*, which means that one either does not know how or is unable to eat a particular food. Some informants were able to describe specific consequences for eating the feared foods, but others would only affirm that they were afraid to eat it or did not know how to eat it. Food taboos were often broken and used as a rationale to explain unfortunate events that occurred among them.

Some differences in food taboos exist among Guajá families. The differences likely relate to the traditional Guajá foraging economy as well as to the formation of the village by FUNAI. Family bands, separated from other Guajá for extended periods of time, may be more likely to undergo cultural drift than sedentary groups who have more consistent contact with other members of the group. In addition, as previously described, some families on the Caru reserve were brought in from other areas as their traditional territory was taken over. While some differences do exist among families, these are not specifically linked to kinship, such as might be seen in totemism.

Five food types were consistently considered to be taboo for all Guajá at all times: capybaras, vultures, bats, large opossums, and most species of snakes. Three species of snakes may be eaten by some: the boa constrictor, the anaconda, and the bushmaster. The boa constrictor is eaten by all the Guajá, but the anaconda and bushmaster are taboo in some cases. In the case of the opos-

TABLE 3.1 Guajá Food Taboos

Food	All Guajá	All Women	Young Women	Nulliparous Women	Young Men
Mammals					
capybara	X	X	X	X	X
large opossum	X	X	X	X	X
small opossum			X	X	
rabbit	X	X	X	X	X
mice	X	X	X	X	
jaguar			X	X	X
tayra			X	X	
bats	X	X	X	X	
river otter			X	X	
squirrel			X	X	
two-toed sloth		X	X	X	X
anteater			X	X	
coatimundi			X	X	
kinkajou			X	X	
bamboo rat			X	X	
porcupine				X	
Birds					
vultures	X	X	X	X	X
Harpy eagle	X	X	X	X	X
other eagles and hawks			X	X	
kingfisher			X	X	
crested guan			X	X	
toucans			X	X	
parrots			X	X	
owls			X	X	X
swift			X	X	
woodpecker			X	X	
ibis			X	X	
Reptiles					
anaconda			X	X	
bushmaster			X	X	
other snakes (except the boa constrictor)	X	X	X	X	X

(continued)

TABLE 3.1 Guajá Food Taboos *(continued)*

Food	All Guajá	All Women	Young Women	Nulliparous Women	Young Men
Reptiles (continued)					
small tortoises				X	
large tortoises			X	X	
frogs and toads			X	X	
turtle (*yačeriho;* undet. Pelomedusidae sp. 2)			X	X	X
turtle (*kwarape;* undet. Pelomedusidae sp. 3)			X	X	
caiman				X	
lizards			X	X	X
Fish					
large catfish (*Pseudoplatystoma tigrinum*)			X	X	X
sting ray			X	X	X
unidentified fish (*akwana*)			X	X	X
unidentified fish (*makari*)			X	X	X
unidentified fish (*maitawa*)			X	X	
unidentified fish (*mara-aiyā*)			X	X	

sum, large ones are taboo, but smaller ones may be eaten. In addition to the large opossum being *inamahə* (bad smelling), it is also called *aiyā* and metaphorically linked to the nocturnal forest cannibal ghosts (discussed in chapter five). Bats are said to give one a headache if eaten. In most families, the harpy eagle and rabbits were also taboo. None of the Guajá ate grubs or any other insect product except honeycomb larvae. Insect eating has been documented for a number of Amazonian groups.[2] The Guajá do not have any specific prohibition on eating insects; they simply do not seem to consider them food. They feel similarly about hummingbirds (*manimə*); the reason they give for not eating them is not that they are taboo, but that they are too small.

The most restrictive food taboos apply to young women, especially from birth until the approach of the end of their childbearing years. Women are classified as *iari* (grandmothers) beginning in their early thirties. *Iari* have very few food restrictions, although many of them continue to bear children. Certain foods are thought to interfere with fertility. Young women with no children are the most vulnerable and have the most restrictions; fewer apply after a woman has had a child. Very few restrictions remain for women past the age of thirty who have had several children, although they may continue to bear more children. Holmberg found a similar taboo pattern among the Sirionó;, the aged who were past childbearing age were free from the taboos of the younger people. Further, among the Sirionó there were some game types that were only supposed to be eaten by the aged: harpy eagles, anteaters, owl monkeys, and howlers (Holmberg 1985:80.).

Foods that are tabooed by all families for young Guajá women are the jaguar, the three-toed sloth, eagles and hawks, the *yačeriho* turtle, and large tortoises (smaller tortoises of the same species can be eaten). The three-toed sloth is called *aiyã* (cannibal ghost), and it is said to cause abdominal pain if eaten. However, the larger two-toed sloth is eaten by young women. In most families, the following animals are not eaten by young women: the small opossum, the tayra, the otter, the anteater, mice, the bamboo tree rat, the squirrel, *arawɨ-te* catfish (*Pseudoplatystoma tigrinum* but not *Pseudoplatystoma fascitum*), *iwa* (toads and frogs), large *kwarape* turtles, the sting ray, the *makari* fish, the anaconda, the bushmaster, *haira* (kingfisher), crested guans, owls, swifts, the *makaratõ* woodpecker, the ibis, toucans, and parrots. In most families, parrots were taboo for young women, and the consequences of eating them were said to be a fever or insanity. In one family, young women were allowed to eat parrots in moderation, with the exception of the small *kiripi* parrotlet.

A relationship may exist between tortoises, the divinity Kiripi, and female fertility. One of the three female divinities, Kiripi is the keeper of the tortoises and maintains a number of them as pets. The parrotlet *kiripi* is considered to be a spiritual sibling with the Kiripi tortoise divinity (see chapter seven). Also of interest is Balée's (1985) finding that the Ka'apor prescribe yellow-footed tortoises for menstruating women. Two Guajá informants stated that all game animals were taboo for menstruating women, although they could eat fish, birds, and reptiles that are not otherwise tabooed. While this is not a prescription for the Guajá, tortoises are the most commonly eaten reptile, and the practice may bear a relationship to that of the Ka'apor. For the Guajá, the specific species of the tortoise does not matter, the proscription is on the size of the tortoise. *Yačeriho*, a species of turtle, is strongly taboo and is considered to be particularly dangerous for pregnant women.

Additional food taboos exist for other states relating to fertility. Nulliparous women may not eat the caiman, the porcupine, or any tortoises, regardless of their size. These taboos do not apply to adolescent girls, even if they are menstruating. Rather, they apply to a special category of sexually mature women who have not been pregnant or have been unable to carry a pregnancy to full term. Certain foods are also proscribed by the couvade. In all families, men with young wives or sexual partners may not eat the *yačeriho* turtle. In other families, particularly for young men, the jaguar, *arawɨte* catfish, the river ray, owls, and several small bird species are taboo. The consequence for breaking a taboo is giving birth to a child that is *manaheĩ* (ugly). During the research period, a child was born with a cleft palate. The explanation for the birth defect was that her father had eaten capybara at one time when her mother carried her. In another instance, a mother's eating *arawɨte* catfish was blamed for a leg infection in the young daughter that she was still nursing.

Young male children are subject to most of the same food restrictions as young girls and women. These apply to boys up to about six years of age.The foods are not so much considered strictly taboo for boys, as they are thought of as foods that the children are not yet developmentally ready to eat. It is understood that eventually they will be able to eat the taboo foods. One unique food restriction for young boys is the tamarin, which, if eaten, will make a boy cry. This is the only restriction on monkey eating outside of the menstrual taboos reported by some women.

Several foods are considered to have special properties if eaten. Young boys are encouraged to eat the jaguar because it is believed to make them strong. Once a boy is past adolescence, he is considered to be fully developed and should not eat jaguar until he reaches the grandfather category, *tamə*, which begins in about the mid-thirties for men. Older men and older women will eat the jaguar for strength, although it is taboo for women, including young girls, until they reach the *iari* stage. The kinkajou is believed to encourage the development of large muscles in men. Small yellow-footed tortoises (*Geochelone denticulata*) are believed to be particularly beneficial for pregnant women, and this belief is therefore similar to the food prescription found among the Ka'apor. No special properties of monkeys could be ascertained. They were considered, however, to be the favorite type of game food, particularly howlers and the capuchins.

DIETARY ASSESSMENT

Two methods were used to assess the relative importance of primates in the diet: random spot-checks and weighing the animals killed for consumption.[3] Both took place during a seven month period from February through August of 1997, which spans the late wet season through the early dry season. The original purpose of the random spot-checks was to provide quantifiable information regarding the types of and time spent in daily activities by the Guajá and the time spent on and types of interactions between the Guajá and their monkey pets. Random spot-checking was adapted from the method termed "instantaneous scan sampling" in nonhuman primate ethology, which records the location and activity of each nonhuman primate at a given time (Altmann 1974). When applied to human groups, the method involves recording the activities that each member of the group is engaged in at a given time and has been used in several studies of South American Indians to quantify the relative amount of time spent in various activities (e.g., Balée 1994a; Gross 1984; Gross et al. 1979; Hill et al. 1984; Hurtado et al. 1985; Johnson 1975).

Randomly spot-checking here involved making one daily round of the village area, including each of the ten Guajá huts. The random spot-checks involved 111 sampling days for a total of 5,206 observations. Thus, each sampling day involved one random spot-check for each individual. The hours for sampling were randomly selected in advance and included the hours between 6 A.M. and 6 P.M. As such, the results reflect only daytime activities. Nighttime observations were not included for several reasons. One obvious reason is that it would have entailed entering into huts while the Guajá were sleeping, which would not have been welcomed. Secondly, I was not willing routinely to travel through the jungle at night to reach the huts, because it would have involved some safety risks, especially snakebite.

Although the original purpose of the random spot-checks was to provide information on the general activity patterns of the Guajá, they also proved to be valuable in assessing the Guajá's diet. Each time an individual was observed eating during a scheduled check, the food was identified and recorded. A total of 248 animal eating episodes were recorded, which represent 4.93 percent of all of the random spot-checks performed. In addition to random spot-checks, game returned from Guajá hunts was identified, recorded, and weighed on hanging weight scales.[4] Data from a total of 120 hunting/fishing days were recorded.

The random spot-checking provided a more accurate measure of the foods eaten than weighing the game did, but there were limitations in both methods. Weighing the game was most accurate for the types of animals returned from a day's hunting trip and also provided a means of verifying the species eaten during the random spot-checks. For longer trips, the game that was returned to the village was often already roasted and partially consumed, making it difficult to obtain an accurate weight. This was often the case with the howlers. According to the Guajá, since they have been living in the village settlement the number of howlers living within a day's walking distance had decreased, and they often had to make longer trips to obtain them. Random spot-checks also provided a means to compare plant and animal foods in the diet, at least in terms of the relative frequencies of eating events. Weighing the game animals provided greater information about the range of animal species utilized. For example, small species or infrequently obtained species would be less likely to be observed during a random spot-check.

In addition, I originally intended to identify and measure the foods eaten by household members. The method was difficult to apply, primarily because of the way that meals are prepared, eaten, and shared.[5] Food-sharing is obligatory among the Guajá.[6] If someone asks for food, it is always given; not only from the cooking pot, but even from the bowl out of which one is eating. Meals themselves usually involve a great deal of activity, with people moving in and out and between the huts. An individual may stop in a hut, eat a little or a lot from another's bowl, bring in additional foods to be shared, and then move on to another hut. Monkeys and other pets may be fed directly from the bowl one is eating from; women often allow pets to eat premasticated foods directly from their mouths.

Two clear seasonal patterns were identified in animal consumption between the late wet season (February through mid-May) and the early dry season (mid-May through August). Howlers were the most utilized species during the late wet season, and their exploitation is associated with increased trekking behavior.[7] During the early dry season, increased fishing behavior is associated with increased sedentism and heavy exploitation of a detritus-feeding, migratory characoid fish of the genus *Prochilodus*, that the Guajá call *piraču*. A significant difference was found between the major animal species utilized in the wet and dry season ($X^2_{df = 3} = 63.09$, $p < .001$).[8]

The hunting of howlers is associated with trekking behavior, while increased fishing is associated with increased sedentism. For the purposes of this study, trekking behavior differs from other hunting/fishing/collecting expeditions in that it involves sleeping away from the main village in the forest. Even the more settled families in the village still spent a substantial portion of their time trekking in the forest. Random spot-checks from one such family indicated that its members spent almost five times more time trekking in the late wet season than in the early dry season. The family was composed of two adult males, two adult females, two adolescent females, and four children, who reside in the same hut in the village, work cooperatively, and are related through the adult and adolescent females, who are wives to the two adult males. A total of 1,110 random spot-checks were made of the family members, with 610 occurring during the late wet season and 500 occurring during the early dry season. A significant difference existed in trekking between the two seasons ($X_{df= 1} = 83.79$, $p < .001$). During the late wet season, individuals were camping away from the village only 5.6 percent of the time (n = 28 trekking, n = 472 in the village). During the early dry season, they spent 26.89 percent of the time camping in the forest (n = 164 trekking, n = 446 in the village). Further, all of the late wet-season camping trips involved dislodging children, pets, and all possessions, while the early dry-season trips involved either males alone or a man with his adolescent wife, leaving the children, possessions, and elder wives in the village. These findings are supported by the work of Queiroz and Kipnis (1991), who observed in 1990 and 1991 that 19 percent of the Guajá were absent from the post during the dry season, while 34.5 percent were absent from the post during the wet season. In addition, FUNAI reports on the Guajá from 1990 mention them hunting in the late wet season and returning with many howlers (Damasceno da Silva 1990).

The behavioral ecology of the howlers also likely affects patterns of their exploitation by the Guajá. Among *Alouatta belzebul* in the Atlantic forest, more than twice as much fruit-eating behavior occurs during the wet season (95 percent) than in the dry season (43.9 percent) (Bonvicino 1989). Fruit eating is one of the primary ways that the Guajá identify the location of prey. Forline (1997) also reported the Guajá actively pursuing monkeys during the wet season, particularly howlers, during the fruiting of *Manilkara huberi* trees. These trees, called *aparihu* by the Guajá, also figure importantly in Guajá mythology (see chapter seven). The Trio and the Matsigenka also hunt monkeys in the wet season, corresponding to the seasonal fruiting of many trees. According to Rivière (1969:45), one Trio informant "could never mention the rainy season without saying in the same breath that it is when spider monkeys are fat." Shepard (1997, 2002) also reports that the Matsigenka prefer to hunt spider and woolly monkeys in the wet season when the monkeys are fatter from eating fruits. Shepard further describes a pattern of wet-season monkey hunting and dry-season concentration on fish, suggesting that focusing on other resources in the dry season alleviates some of the hunting pressure.

During the early dry season, fishing dramatically increases in importance, rising from a relative percentage of 9.3 to 44.37 of all random spot-checks. Other aquatic resources are exploited in greater numbers as well, such as turtles, otter, and caiman. The most important fish species is the characoid fish, *piraču* (*Prochilodus* sp.; *curimatá* in Portuguese), which is heavily exploited during its "*piracema*," the migration from the receding waters in the tributaries to the Pindaré river at the start of the dry season (see Goulding 1980 for a description of *Prochilodus* seasonal migration patterns).[9] The diminishing rainfall also facilitates Guajá bow and arrow fishing as water turbulence decreases and the *piraču* are more readily visible. Later, as the waters recede farther, the Guajá exploit a variety of fish trapped in pools by use of fish poisons.[10]

Prochilodus fish are hunted with bow and arrow at night, when they are "sleeping" according to the Guajá. The Guajá on the Caru Reserve typically use flashlights at night when travelling through the forest and use flashlights when fishing at night.[11] They also still use torches fueled by pieces of a hardened tree resin wedged into the split end of a small tree branch (*Manilkara huberi* resin according to Balée, personal communication, 1999). An early report of South American Indians on the Tapajós reported them spear fishing at night using flaming bundles of green bark stripped from palm fronds (Bates 1864, in N. Smith 1981), and the Carajá were described at mid-century as spear-fishing at night (Limpkind 1948, in N. Smith 1981:50–51). The Guajá have recently adopted fishing with line and hooks, and they acquire smaller fish in this manner. Women typically fish with fishing line, but they will on occasion borrow a bow and some arrows from a spouse or male relative to fish. One male also routinely used a fishing net.

The early dry season is also an important ritual time for the Guajá when the *karawárə* ceremony is practiced. In brief, several times each week, at night, the village gathers to sing in a ceremony involving contact with the divinities and dead ancestors who reside in the sky home, the *iwa*. While the men dance, their spirits are transported to the *iwa*, and the divinities' spirits possess the bodies of the dancers on the ground. The divinities are believed to have the power to heal the village and to bring game close to the Guajá. Women and children participate by singing and dancing, but they neither go to the *iwa* nor are spirit-possessed during the ritual.

The *karawárə* ritual is also important in understanding the relationship between seasonal environmental change and Guajá social organization. Marriage arrangements are formed through consultation of the divinities during the *karawárə* ritual. The local, seasonal availability of fish during the early dry season allows the large group to congregate for a sufficient amount of time so that such arrangements can be made. I was present during the final three months of the dry season in 1997, and no first marriages were arranged, but seven new first marriages were arranged at the beginning of the dry season when the *karawárə* rituals began.

Guajá girls marry at around six or seven years of age, and men perform a type of brideservice which involves bringing food to the family of the girl. Culturally, this is not seen as a form of payment or exchange but is explained as bringing fish to the girl so she will grow and not be afraid to live with her new husband. Although the cultural obligation exists to marry once an agreement has been made between a prospective husband and a girl's father, girls are not expected to leave the natal home until they are no longer afraid to do so. Stories of marriages tell of this period of waiting while the young girl eats fish. No prescription or special properties of fish were ascertained that would help specifically in this process, and informants agreed that the girl would be eating many other things as well, but "fish eating" is the way that this marital transition is described.[12]

White-lipped peccaries are also an important food source for the Guajá. When combining the late wet and early dry seasons, peccaries were the most utilized species overall, although they were secondary to howlers in the wet season and fish in the dry season. Peccary hunting in general is more opportunistic. When evidence of a herd of white-lipped peccaries is spotted in the area, the Guajá literally drop everything and take off running to the forest. Although the Guajá keep dogs, they are not used in hunting white-lipped peccaries.

Plant foods also play an important role in the diet. In the observed eating episodes, plant foods were eaten much more frequently than animal foods. In addition, the Guajá on the Caru reserve increasingly rely on domesticated plants, which is indicative of their current shift from a foraging mode of production to a more sedentary, horticultural mode of production. Manioc, in particular, is becoming an increasingly important caloric staple and is perhaps replacing the babassu palm as the staple plant food.

Analysis of eating episodes is limited in that it cannot provide information on the quantity of food eaten at each episode, which makes analysis of the precise nutrient composition of the diet problematic. However, it does provide a general picture of the prevalence of food types and a general description of the diet. Over twice as many plant eating episodes occurred compared to animal eating episodes: 67.55 percent (n = 537) plants to 31.19 percent (n = 248) animals with 1.26 percent (n = 10) honey. Of the plant eating episodes, 76.54 percent (n = 411) were of domesticates and 23.46 percent (n = 126) were of nondomesticates. If it is assumed that the Guajá may choose whether to consume domesticates or nondomesticates, a chi-square goodness-of-fit test demonstrates a significant difference ($X^2_{df = 1}$ = 151.25, p < .001) between these types.

In comparing nondomesticated foods (plants, animals, and honey) with domesticated foods, little difference exists between the two categories: 51.70 percent eating episodes were domesticates, and 48.42 percent were non-domesticates. What is notable, however, is not the differences between the categories. Rather, the data suggest that more than half of the Guajá diet derives from domesticates, making it clear that the Guajá are a culture in transition from foraging to horticulture.

Of particular importance is the role of manioc in the diet. Bitter manioc accounted for 43.76 percent of all plant-eating episodes and 29.56 percent of the total number of eating episodes. Including both bitter and sweet varieties, manioc accounted for 50.09 percent of plant-eating episodes and 33.84 percent of all eating episodes. The traditional staple of the Guajá, babassu, accounted for only 4.28 percent of the plant foods eaten and 2.89 percent of the total diet. It would seem that manioc is replacing babassu as the staple food of the Guajá.

GUAJÁ ETHNOBOTANY

Originally, in order to assess the relative importance of primates in the diet of the Guajá, plans were made to collect and identify the edible plants available to the Guajá. Plants edible by humans known to the Guajá were few and widely scattered. However, informants pointed out many plants eaten by game on plant-collecting trips. At this point, I decided to do a more extensive plant collection in order to evaluate the types of knowledge the Guajá possessed about their environment and, specifically, the relative importance of their knowledge of the plants eaten by monkeys. The results here demonstrate that Guajá ethnobotany is heavily monkey-oriented, in that knowledge of plants eaten by monkeys is key to understanding Guajá perception of the forest flora.

Such an assessment does present a problem in terms of Balée's considerable evidence that the Guajá have experienced agricultural regression (see Balée 1994a, 1996, and 1999). Specifically, the evidence presented here suggests that Guajá ethnobotanical knowledge is functionally integrated with a foraging mode of production that concentrates on monkeys as game. Balée's historical ecological evidence suggests otherwise: that the Guajá mode of production represents cultural loss from a previous horticultural existence. However, monkey-oriented plant knowledge and agricultural regression need not be mutually exclusive explanations. Below, selective attrition is presented as a possible means to integrate these two lines of evidence.

Identifications through the Museu Goeldi and the Tulane University Herbarium have differentiated my plant collections into 275 useful, noncultivated species and morpho-species.[13] By far, the most prevalent knowledge of the botanical environment involved plants that were eaten by game. While 84 percent of the plants had at least one use that included being eaten by game, only 14.91 percent were eaten by human beings. Further, knowledge of what monkeys eat is predominant among game animals, as over half of the plants identified as being eaten by game were eaten by monkeys (51.94 percent).[14, 15]

Determining the statistical significance of Guajá knowledge of plants eaten by monkeys is somewhat problematic since approximately 25 percent of the plants have multiple uses (n = 64). Some plants may also be eaten by one or more nonprimate species, or the Guajá may possess other important knowledge of the plant, such as its having medicinal or magical value. Thus, creation of mutually exclusive categories of Guajá plant knowledge is difficult. However, if an assumption is made, for purposes of statistical analysis, that the Guajá would know as many plants that are eaten by nonprimate mammals as they do plants that are eaten by monkeys, a chi-square goodness-of-fit test demonstrates significant differences ($X^2_{df=1}$ = 17.04, p < .001) between the two categories. In terms of plants that the Guajá know are eaten by mammals, 65.21 percent are eaten by monkeys and 34.78 percent are eaten by other mammals (N = 184 total plants eaten by mammals).

The implication of the data is that, in terms of breadth of knowledge, it seems more valuable for the Guajá to know the plants that game animals eat than it is to know the plants that they themselves can eat. One might be tempted to characterize the Guajá as more of a "hunting" people than a "gathering" people. A similar assessment was made of the hunting and gathering Maku of the Northwest Amazon; Silverwood-Cope (1972) has described them as "professional hunters" who rely on trading with Tukanoans for domesticated plants to gain access to sufficient carbohydrates (also see Milton 1984). However, as previously noted, the random spot-checks indicated that the Guajá utilize significantly more plant resources in their diet than animal resources. The emerging picture seems to be that the Guajá rely on a few reliable plant resources (babassu in the past, which is being increasingly replaced with domesticated manioc) with the *diversity* of their plant knowledge geared towards plants that are eaten by game.

The botanical data give a great deal of insight into how the Guajá perceive and utilize their environment. However, the botanical data are not meant to suggest that Guajá's knowledge of the diversity of plant life of a type of prey corresponds directly to Guajá's intensity of utilization of a type of prey for food. One obvious reason is that not all of the Guajá's prey are strictly herbivorous. The more omnivorous species are underrepresented in the sample, and some carnivores that the Guajá eat, such as the jaguar and caiman, are not represented at all. In addition, dietary breadth is variable among species, and Guajá knowledge of dietary diversity would reflect, to some extent, the inherent diversity (or lack of diversity) in the diet of a given species.

In addition, species' behavioral patterns may affect Guajá knowledge of plant utilization. The feeding habits of nocturnal, arboreal, or solitary-living species would be more difficult to observe than that of diurnal, terrestrial, or group-living species. The utilization of calls by a species may affect Guajá knowledge of plant utilization. For example, howlers advertise their presence with loud territorial calls, which makes locating them and recognizing their eating habits easier. On the other hand, such advertisement may make knowledge of feeding behavior less important. Migratory white-lipped peccaries travel in herds and may be readily seen or heard when in the area, necessitating less reliance on indirect evidence of eating habits to locate them. An additional problem is that some species are exploited for reasons other than food. Toucans, for example, are eaten, but are also sought for their feathers, which are used as body decoration in the *karawárə* ritual. Vultures also are sought for their feathers, but they are never eaten.

Despite those caveats, the botanical evidence suggests that the diversity of Guajá plant knowledge is oriented toward hunting rather than gathering, since they know far more about what is eaten by prey than what is edible for themselves. Knowledge of the plants eaten by game is an important hunting strategy for locating animals in the forest. According to Lizarralde (2002), the most successful hunters of spider monkeys among the Barí had extensive knowledge of their feeding trees. Further, the evidence clearly points to monkeys' being particularly important prey, since the Guajá devote much of their ethnobotanical ken to the diet of monkeys.

Balée has compared the ethnobotany of the Ka'apor and the Guajá Indians, who both speak closely related Tupí-Guaraní languages and exploit the same environment. However, the Ka'apor have an agricultural mode of production, while the Guajá have a foraging mode. Balée argues that historical factors have resulted in Guajá cultural loss and regression from a previous horticultural mode of production. He provides four strong lines of evidence in support of this: Tupí-Guaraní cognates with domesticated plants in the Guajá language; Guajá terms for nondomesticates modeled on terms for domesticates; fewer plant names known by the Guajá than the Ka'apor; and underdifferentiation of generic plant names by the Guajá in comparison with the Ka'apor.

Balée (1999:28) has described a number of cognates existing in the Guajá language for domesticated plants such as *kara* (yams), *kamana'ĩ* (beans), *kwi* (calabash tree), *paku*[16] (banana), *wači* (corn), and *waya* (guava). Two possible explanations for the existence of the similarity in these terms between the languages are that they reflect traditional knowledge of domesticates that has been lost or that they are not cognates at all, but rather the result of cultural borrowing from other Tupí-Guaraní speakers. Balée (1999:29) argues that linguistic borrowing is unlikely due to the violent nature of the interactions between the Guajá and other horticultural groups. I am less inclined to dismiss the possibility of linguistic borrowing completely. Although the characterization of Guajá relations with neighbors as hostile is apt, it is certainly possible that they have had some sort of cooperative interaction at some point with some group during their culture history as hunter-gatherers. And even if this is not the case, hostile interactions can lead to kidnapping, which could have presented a mechanism for cultural exchange. For example, the Guajá report that the Guajajara used to kidnap their wives and children, and an escape could have presented the opportunity for linguistic borrowing.[17]

However, while I would generally consider linguistic borrowing as a stronger possibility than Balée does, it is clearly both insufficient and inadequate to account for all the features of Guajá ethnobotany and culture that suggest horticultural regression. Perhaps Balée's (1994a, 1999) strongest evidence comes from the Guajá's modelling names for nondomesticates on the names for domesticates, which he argues suggests the historical importance of the domesticate. For example, domesticated *Bixa orellana* is called *araku-ate* ("true" *araku*) in Guajá while nondomesticated *Bixa arborea* is called *araku-rána* ("false" *araku*) (Balée 1999). Another example involves the Guajá term for a nondomesticated manioc species (*Manihot* sp.), which is a marked Tupí-Guaraní cognate: *arapʰa-mani'ɨ* (brocket deer manioc) (Balée 1999: 29). However, the term for domesticated manioc is *tarəma*. Here, the marked nondomesticate lacks an unmarked domesticated counterpart, which points to a cultural loss of the domesticated species. Interestingly, the Guajá refer to a mythical people who live underground called the *Manio'i*, a term that also utilizes the Tupí-Guaraní manioc cognate. The metaphorical association of an underground people with an underground root crop suggests that manioc was a culturally significant plant at one time in Guajá history.

Balée also describes Guajá ethnobotanical knowledge as simplified in comparison with the Ka'apor. According to Balée (1999:35), the total number of Ka'apor generic plant names is 40 percent higher than that of the Guajá. Not only are there fewer plant names overall, but the Guajá also demonstrate less differentiation into folk-specific categories. What is most striking is that this holds true even when just comparing nondomesticated plants. The expectation would be that the Guajá would have more elaboration of nondomesticates and the Ka'apor would have more elaboration of domesticates, which would be consistent with their corresponding modes of production. However, the Ka'apor have five times the number of nondomesticated folk specifics than the Guajá (Balée 1999:38). The ethnobotanical impoverishment here is suggestive of cultural loss.

Even more dramatic evidence for general cultural loss is that the Guajá on the Caru reserve do not know how to make fire. Fires are kept burning and started from firebrands, but they do not use the bow drill or other technology to make fire. Given the importance of fire to all human cultures, it seems undeniable that the Guajá must have experienced cultural regression to have lost this important technology. Arguably, fire-making ability would be an even more important technology for hunter-gatherers than it would be for horticulturalists. Should a fire go out, small family bands may not have easy access to another group to get a firebrand. In fact, informants complained of this very problem in the wet season with heavy rains putting out their fires while they were away in the forest. With firemaking being virtually a human cultural universal, such a loss clearly suggests that the Guajá have experienced serious upheaval in their culture history.[18]

The evidence that the Guajá have regressed from a horticultural past does not necessarily refute the evidence for the functional integration of monkey exploitation in the more recent Guajá hunting-gathering-fishing mode of production. One way to examine the compatibility of the two explanations is to determine whether Guajá cultural loss is random or nonrandom. Random loss would suggest that Guajá ethnobotany can be considered a remnant from a former, more extensive ethnobotanical lore. On the other hand, nonrandom loss would suggest that a process of cultural selection made the Guajá more likely to retain certain types of plant knowledge. In other words, the change to a foraging mode of production coincided with the loss of plant knowledge least useful to foraging and the retention of plant knowledge most useful to foraging. The evidence suggests that cultural selection has taken place, as well as some cultural innovation.

Random attrition of plant names would indicate the indiscriminate loss of plant knowledge. Attrition of plant names can be considered analogous to the bottleneck effect in genetic drift. The Guajá may have suffered depopulation through disease and warfare during the early European colonization with only a subset of the collective ancestral knowledge of the group being retained. The assumption of random attrition would be that no difference exists in the types of knowledge that were retained and the types that were lost.

Nonrandom loss suggests that while loss of plant knowledge occurred, it occurred through selective enculturation of a subset of the ancestral knowledge. Here, factors in Guajá history may have dislocated and reduced their numbers, but plant knowledge most useful to foraging was retained, while knowledge most important to horticulture alone was lost. In this scenario, the shift to foraging would have entailed emphasizing certain types of knowledge over other types of knowledge, and the most relevant subset of ancestral knowledge was transmitted from parents to children. Two lines of evidence suggest such cultural selection: differential retention of names of domesticates and differential retention of plants eaten by game.

Balée (see 1994a and 1996) has noted that less settled, trekking horticultural groups in the Amazon rely on maize more heavily than manioc because of the relatively low start-up costs of maize in comparison with the slow-growing, extensive processing needed for manioc. This seems consistent with evidence from the Guajá. As previously mentioned, Tupí-Guaraní cognates for bananas (*maku* or *paku*) as well as for corn (*wači*) are present in the Guajá language, but not for bitter manioc (*tərəma*). This seems to suggest selective attrition involving either sequential loss of domesticates or perhaps retention of the names of crops that could be stolen easily. Manioc is harder to steal than corn or bananas because it has to be dug from the ground. Bananas and corn may have been retained because they remained part of the diet, through crop-stealing, while manioc did not.

In terms of nondomesticates, differences in plant knowledge between the Ka'apor and Guajá suggest that the Guajá have selectively retained plant knowledge most useful for hunting. Balée describes four categories of nondomesticated plants used by both the Ka'apor and the Guajá: tree, vine, herb, and unclassified nondomesticates. Among the Ka'apor, a significant difference ($X^2_{df = 3}$ = 81.59, p < .001) exists between the types of botanical plant species known to be eaten by game animals and those that are not eaten by game animals. However, when comparing just the categories of herb, vine, and unclassified nondomesticate, no significant differences exist among the categories. Thus, the difference exists among trees, in that 72 percent of the botanical species of trees known by the Ka'apor are eaten by game animals. Balée found, in comparing Guajá and Ka'apor plant categories, that while the Ka'apor knew 486 percent (41 versus 7) more plant names for nonclassified nondomesticates and 62 percent (60 versus 37) more names for vines, they only knew 20 percent (257 versus 214) more names for trees. If the assumption is made that the Guajá and the Ka'apor had similar types of botanical knowledge as horticulturalists, then it would seem that the Guajá lost knowledge of fruiting trees heavily exploited by game at a slower rate than they did of vines and nonclassified nondomesticates that are less heavily exploited by game.

It should be mentioned, however, that the Guajá actually know 9 percent (63 versus 58) more names of herbs than the Ka'apor. While the reason for their more extensive knowledge of herbs is unclear, it does not refute the evidence of selective attrition. In fact, it supports it. However, the question remains as to why they know more names of herbs that the Ka'apor. One possible explanation is that the leaves of many herbs are believed to have magical properties and are crushed and used in medicinal baths. The easy accessibility of the leaves of herbs, particularly in comparison with trees, may help explain why the Guajá use them in this way.

Another method of determining differences betwen the Guajá and Ka'apor ethnobotanies is to compare the relative number of plants eaten by game versus those not eaten by game for both groups. Since Balée excluded plants eaten by birds in his table of animal foods, those eaten by birds and no other animal were excluded from my data with the Guajá. A significant difference ($X^2_{df = 1}$ = 15.13, p < .001) exists between the relative number of plants eaten by game known to the Guajá in comparison to the Ka'apor (Guajá: 203 game plants versus 72 nongame plants; Ka'apor: 417 game plants versus 272 nongame plants).

In addition, the anthropogenic habitats exploited by the Guajá may also attract monkeys. Balée (1994a:133) identified thirty-four species of plants that are dominant colonizers of old fallow forests. Of these old fallow species, the Guajá described approximately 35 percent of them as plants eaten by monkeys (see Cormier 2002). Thus, monkey hunting may be interrelated with Guajá adaptation to the plants in old fallow forests.

In conclusion, the ethnobotanical evidence suggests that the Guajá have both undergone cultural loss and a process of reorientation toward a foraging lifestyle. Loss of the knowledge of plants seems to have taken place, but the evidence suggests that the process was nonrandom and involved cultural selection. Cultural selection here does not necessarily imply choice. That is, I am not suggesting that the Guajá chose to be foragers rather than horticulturalists. Rather, I am suggesting that the change to a foraging lifestyle was likely to due to historical factors such as devastation through disease, warfare, and depopulation, which quickly and radically altered the cultures of many indigenous New World peoples, even driving some to extinction. While the Guajá may not have chosen foraging, once in a situation where horticulture was not possible, the evidence suggests that certain types of knowledge were retained over others.

4

GUAJÁ KINSHIP

Monkeys are an important food source among the Guajá, but they are also considered to be fundamentally kin. In order to explain further the social role of monkeys among the Guajá, a general treatment of their kinship system is needed. Below is a description of the Guajá kinship terminological system, marriage patterns and sexual behaviors, the role of consanguinity and affinity in reckoning kinship relations, genealogical amnesia, and the functions of ambiguous classification of relations.

THE DRAVIDIANATE-AVUNCULATE KINSHIP TERMINOLOGICAL SYSTEM

The Guajá are similar to many other lowland South American groups in that they use a Dravidian kinship terminological system, also known as the two-line prescriptive system (Rivière 1984). Although several variants of the Dravidian system have been described in the literature[1] (e.g., Good 1980; Trautmann 1981; Godelier, Trautmann, and Tjon Sie Fat 1998), common key features include bifurcate merging, distinguishing cross-cousins from parallel cousins, merging parallel cousins with siblings, and the encoding of affinal terms. The Guajá version of the Dravidianate encodes spousal terms for the avunculate or oblique

---

FIGURE 4.1   Child with tamarin (*Saguinas midas niger*)

marriage. The avunculate involves the marriage of a female ego to her mother's brother and is a pattern also found among other Amazonian groups, such as the Yuquí (Stearman 1989), the Parakanã (Fausto 1997), and the Trio (Rivière 1969). In the discussion below, abbreviated kinship references follow the conventional anthropological system of notation: F = father; M = mother; D = daughter; B = brother; Z = sister; H = husband; W = wife.

The spousal term is encoded in the Guajá kinship terminological system so that a female ego calls her mother's brother *iménə* (husband), and a male ego calls his sister's daughter *emeriko* (wife). A male ego also calls his MBD by the *wife* term because when an avunculate marriage is practiced, MBD is simultaneously ZD. Several key formulas can be described for the encoded avunculate marriage (see table 4.1):

MALE EGO	FEMALE EGO
M = FZD	M = FZD
MB = FZS	MB = FZS = husband
ZD = MBD = Wife	S = MBS; D = MBD

Guajá marriage patterns conform to their avunculate terminology. A principle of generational extension of affinity exists, whereby the ZD term may be extended to ZDD (if not ego's daughter) and ZDDD. It is also reflected in a female ego's extension of the term MB to MBS (if not ego's son) and MBSS. Hornborg (1998:172) describes this type of construction as the "parallel transmission of affinal relationship" drawing on Dumont's (1953) early description of the inherited alliance relationship found in Dravidian systems. Taylor's (2001) description of consanguinity conceived as a kind of cloning among the Jivaro may represent a more apt way to view this than as inheritance. Here, from ego's perspective, the consanguineal relationships of affines can be considered to be replicated from one generation to the next. It should also be mentioned that parallel transmission creates ambiguity in the system, allowing relatives to be classified in more than one way. For example, Guajá girls and women may call their MBS either "husband" or "son."

The precise kinship relationship between two married individuals was frequently difficult to trace, which is not an uncommon problem for anthropologists working in Amazonia due to the prevalence of genealogical amnesia and other related factors that tend to deemphasize descent. For example, Rivière (1969) was only able to identify the genealogical links for ten marriages among the Carib-speaking Trio, despite having spoken to scores of people.

TABLE 4.1 Terms of Reference and Address

Category	Male Term of Reference	Male Term of Address	Female Term of Reference	Female Term of Address
FF, MF	tamə	čipa-te, čipa-tamə	tamə	čipa-te, čipa-tamə
FM, MM	iari	amā-čia-te	iari	amā-čia-te
F	tu	čipa	tu	čipa
FB	tu-na, tunawana	čipa	tu-na	čipa
M	ɨhi	amā	ɨhi	amā
MZ	ɨhi-na	amā	ɨhi-na	amā
FZ	yahɨ	amā or čikari	harəwaiya	čiá
MB	harəwaiya	čiá	iménə	čipa'i or personal name
B	harəpihárə	harə	ikuwɨčiá	či
Z	haininawai	čikari	harəpihárə	harə
FBS	harəpihárə	harə	ikuwɨčiá	či
FBD	haininawai	čikari	harəpihárə	harə
MZS	harəpiana	atɨ, harə	ikuwɨčiá	či
MZD	haininawai	čikari	harəpihárə	harə
FZS	harəwaiya	čiá	iménə	
FZD	ɨhi-na or emeriko	āma, čikari, čiku or čia'ə	ɨhi-na	amā or harə
MBS	harəwaiya	čiá	mɨmɨrə or iménə	tači, čipa'i, or personal name
MBD	emeriko	čiku or čia'ə	mɨmɨrə	
S	mɨmɨrə, ta'ira	atɨ	mɨmɨrə	tači
D	mɨmɨrə, ta'ira	čikari	mɨmɨrə	čikari
BS	mɨmɨrə	atɨ	ipenčiá	či
BD	mɨmɨrə	čikari	harəwaiya	čiá
ZS	harəwaiya	čiá	mɨmɨrə	tači
ZD	emeriko	čiku or čia'ə	mɨmɨrə	čikari
GS	haminaminu, harəwaminu, harəiraira	personal name	hanimiaru	personal name
GD	haminaminu, harəwaminu	personal name	hanimiaru	personal name

In all, fifty-five marriages and betrothals among the living Guajá were identified, which included past marriages between living individuals who have since divorced. Of those fifty-five, sufficient genealogical information was gathered to determine a kinship link for twenty-seven of them, in part due to childhood marriages of girls (see below) whose parents were still living. Of the remaining marriages, a kinship link could not be determined for twenty-two marriages, and six marriages involved the incorporation of outsiders who were not considered by the Guajá to have a direct a kinship link with the group. Thus, excluding marriages to unrelated individuals, slightly more than half of the genealogical relationships could be determined. For that reason, the frequencies described below cannot be taken to reflect the actual frequencies of marriage types, but rather, the frequencies of marriage types among those where a kinship link could be determined.

Four marriage patterns were identified, each of which is, at least terminologically, an oblique marriage, that is, not in ego's generation. The most common form was the avunculate, or ZD marriage, followed by ZDD marriage (= WD marriage), ZDDD marriage (= WBD marriage), and the "M" marriage (= FZD marriage) (see Figure 4.2).

The avunculate type involves the marriage of a male ego to his sister's daughter, or of a female ego to her mother's brother. There were ten marriages and betrothals of this type, representing approximately 37 percent of all the marriage links identified and making it the most common form identified. In some groups that practice the avunculate, a distinction is made between a male ego's older sister and his younger sister, allowing or preferring marriage to the older sister's daughter, but prohibiting marriage to the younger sister's daughter (e.g., the Trio [Rivière 1969] and the Parakanã [Fausto 1997]). No terminological distinction could be ascertained among the Guajá between the older sister and the younger sister, and marriages occurred with both types of sisters. However, older ZD marriages or betrothals were more common, which is probably the most logical outcome, since an older sister would likely be the first sister to produce a daughter eligible for marriage.

It should also be mentioned that Erikson (in press) has described the adoption of the avunculate marriage among the Matis of Amazonas as a result of epidemic disease. The Matis have a *kariera* system that forbids oblique marriage. According to Erikson, oblique relatives are being reclassified into marriageable categories due to the shortage of marriage partners following depopulation by disease. While the avunculate marriage is both preferred and encoded in Guajá kinship terminology, Erikson's findings demonstrate that the oblique marriage is a response to historical events in at least one group.

Eight cases of WD (or ZDD) marriage were recorded, representing approximately 26 percent of those identified, and it was the second most common type identified. Typically, this involved marriage to the daughter of a classificatory wife, but in one instance, a man was in a polygynous marriage that involved simultaneous marriage to two women who were mother and daughter to each other. The daughter was fathered by another man, so the husband was not her biological father (see below for Guajá ideas regarding paternity). From what history could be reconstructed, this may have been a pattern in this family. The elder wife had been previously married to her husband's father. It was unclear, however, whether the elder wife had been serially or simultaneously married to the father and son.

While this case represented the only actual simultaneous marriage to a mother-daughter pair seen, there were other cases of mothers and daughters sharing sexual partners. In one household where a mother-daughter pair resided but were married to different spouses, the mother's husband fathered several of the mother's daughter's children, according to the cultural belief that all male sexual partners during a pregnancy are "biological" fathers (to be discussed below). According to Rivière, marriage of a man to a mother-daughter pair is a widely reported practice among Carib-speaking Indians, and Trio history recorded two such marriages (see Rivière 1969:161). In addition, he reports two cases of women serially married to a father and a son (Rivière 1969:160). Wife's daughter (ZD) marriage has also been reported among the Jivaroan Shiwiar (Seymour-Smith 1991: 639).

Included in the ZDD/WD marriage were two cases of ZHZD marriage. Here, a male ego marries his FZDD. Arguably, the arrangement could be called "sister marriage" or "mother's daughter's marriage," but this would be a culturally inconsistent interpretation because mother's daughter (any sister) is strictly taboo. The best interpretation of this form is perhaps as a variation on the WD marriage, when the patrilateral cross-cousin (a classificatory mother) is alternatively classified as "wife" (as discussed below in the fourth type of marriage). Two of the marriages observed involved the relationship of ZHZD.

Seven cases of the ZDDD type were recorded, representing approximately 24 percent of those identified. Here, a man marries his wife's brother's daughter, who is literally or terminologically his wife's granddaughter or his sister's daughter's daughter's daughter. In two of the marriages, a man was simultaneously married to both a woman and her granddaughter. In both cases, the man was first married to an older woman and later married her granddaughter, creating a polygynous union of the grandmother and granddaughter to a single husband. In neither case was the man considered to be a biological father to the elder wife's daughter. Thus, the granddaughter was not considered to be his own biological granddaughter.

The WBD marriage may also function in combination with the avunculate as a type of bilateral daughter exchange between ego and ego's sister's husband (ego and his brother-in-law). One direct exchange of this type was negotiated during

FIGURE 4.2 Kinship charts

a. ZD Marriage: The Avunculate marriage

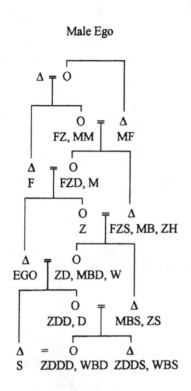

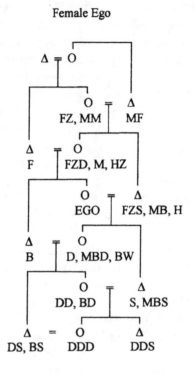

Male Ego

Female Ego

b. ZDD "Wife's" Daughter Marriage

a. ZDD Marriage

b. Polygynous WD Marriage

c. ZHZD/FZDD as "Wife's" Daughter Marriage

ZHZD Marriage

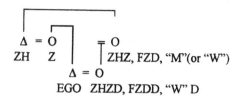

FZDD Marriage

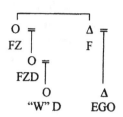

d. ZDDD: WBD or "Wife's" Granddaughter Marriage

WBD Marriage

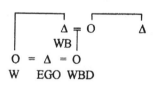

Polygynous ZD and ZDDD Marriage

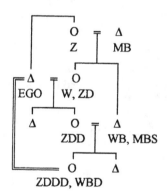

e. Bilateral Daughter Exchange

the study period. The informant described trading his daughter to his wife's brother in exchange for the wife's brother's daughter. Both of these were betrothals, and the daughters remained in their respective natal households. The daughter of the informant was about seven years old and would likely soon begin the marriage transition. The daughter of his wife's brother was still a toddler, so several more years would pass before she would begin the marriage transition.

Two marriages were to the patrilateral cross-cousin. The patrilateral cross-cousin type involves the marriage of a male ego to his father's sister's daughter, who is a classificatory "mother." Since the patrilateral cross-cousin is a classificatory mother, this form is at least terminologically oblique. One involved a man's marriage to an actual father's sister's daughter, while the other involved the marriage of a man to the daughter of a classificatory father's sister. The classificatory "mother" marriage is not unique to the Guajá. Among the Suyá, as previously mentioned, the preferred marriage is to the matrilateral cross-cousin, who is a classificatory mother in their Omaha system (Seeger 1981). Among the Trio, marriage to a classificatory mother (nosi) who is MMD or FZD is both prescribed and the most popular marriage form (Rivière 1969:67–68). No one thinks, however, that they are marrying their birth mothers.

In addition, another common marriage pattern was the marriage of a man in his teens or twenties to a woman who was considerably older. The first wife of a young man is sometimes an older woman whose husband has died. Unfortunately, I was not able to trace the genealogical ties in any of these marriages.

SEX, MARRIAGE, AND PLURAL PATERNITY

Plural Paternity

The Guajá reckon kinship paternally, however, paternity itself is a plural concept due to their notions of what has been called partible paternity.[2] Partible or plural paternity is a widespread concept in Amazonia and has been documented in numerous language groups such as the Gê-speaking Suyá (Seeger 1981), the Arawakan-speaking Mehinacu (Gregor 1974), the Chapacuran-speaking Wari' (Conklin 2001, Vilaça 1992), the Panoan-speaking Matis (Erikson in press), and the Tupí-Guaraní-speaking Tapirapé (Wagley 1983). The basic belief is that the fetus develops from the accumulation of semen. Therefore, it is possible for a child to have more than one "biological" father.

Among the Guajá, plural paternity is not viewed as merely a potential for multiple genitors, but more as a necessity in that the amount of semen needed to create a child is viewed as more than one man alone would normally be able to provide. Erikson (in press) describes extramarital sexual relationships as being an almost mandatory social responsibility for the Matis, who also have the notion of plural paternity. Guajá men feel a tremendous obligation, and at times it is expressed as a burden, to have frequent sex with their pregnant wives in order to meet the semen demands for fetal development. Women seek out sexual partners in addition to their husbands (or sometimes instead of their husbands), and the modal number for any given pregnancy is three. Thus, while individuals have only one birth mother, they have several men that they consider to be their true, "biological" fathers, because each has contributed semen to create them. These biological fathers are any man who has sexual relations with the mother either just prior to or during a pregnancy. The Guajá believe that females have no role in conception or fetal development apart from serving as a kind of container to house the semen men provide.[3]

Despite the belief in paternally linked consanguines, it would be inaccurate to classify the Guajá's system as patrilineal. When an individual has multiple fathers who have multiple fathers, there is no clear-cut "line." Erikson (in press) also makes the important point that partible paternity weakens the role of paternity, as the category of *father* becomes a more generic classification. The children of a man are considered to be his consanguines, as are brothers and sisters linked through the same father. The patrilateral parallel cousins also receive the consanguineal sibling term. Although the Guajá have the notion of patrilateral consanguinity, it is not expressed in terms of descent groups, which is characteristic of unilineal systems. All kinship reckoning is egocentric. It is uncommon for two individuals to share the same set of fathers because mothers change sexual partners frequently (there were no cases determined among the Guajá where any individual shared the same father set with another individual). Although two individuals may be *harəpihárə* siblings to each other, their other *harəpihárə* siblings may not stand in this relationship to each other. Thus, the *harəpihárə* group varies from individual to individual.

Plural Marriage

Guajá marriage practices employ both polygyny and polyandry, remarkably, sometimes at the same time within the same marriage. The polgynous forms are more common on the Caru Reserve than the polyandrous forms. On the Caru reserve, polyandry is rare and best characterized as a transitional state between marriage partners to effect a divorce.

Marriage patterns are different for men and women due to the oblique marriage practice. While women tend to have their husbands sequentially, men often have two wives simultaneously. Since girls marry at around age six or seven to a husband who may be significantly older than they are, by the time they are in their mid-twenties, the first husband has often died, and the wives re-marry. Many times the second husband is a younger man. For a man, he may have both a wife that is much younger than he is as well as a wife that is around his age, or perhaps considerably older. Men who marry young girls take on a parental role (to be discussed further below). The same is true for the husband's elder wife who often takes on a more maternal role to the young co-wife. Frequently, co-wives are related as sisters or classificatory sisters, but sometimes, as described above, grandmother-granddaughter and mother-daughter polygynous arrangements occur.

As mentioned above, there were some cases of plural type marriages where both polygyny and polyandry were practiced. An example is illustrated below.

$$\underset{Maihu}{O} = \underset{Čiami}{\Delta} = \underset{Iname}{O} = \underset{Takamači'a}{\Delta} = \underset{Ipiõčika}{O}$$

In this household, Čiami had two wives, Maihu and Iname. Iname was also married to Takamači'a who had another wife, Ipiõčika. However, this was not a group marriage. Takamači'a was not married to Maihu, who was his consanguineal sister, and Čiami was not married to Ipiõčika, who was a classificatory daughter. In this case, Čiami was first married only to Maihu, and Takamači'a had two wives, Iname and Ipiõčika. All were living in the same household and Čiami and Iname were sexual partners. When Iname became pregnant, Čiami began referring to Iname as a true wife. At this time, Iname was also considered the true wife of Takamači'a. Some time later, Takamači'a discontinued referring to Iname as his true wife (ha-meriko), and she became only a classificatory wife. Thus the pattern was:

$$\underset{Maihu}{O} = \underset{Čiami}{\Delta} \qquad \underset{Iname}{O} = \underset{Takamači'a}{\Delta} = \underset{Ipiõčika}{O}$$

$$\Downarrow$$

$$\underset{Maihu}{O} = \underset{Čiami}{\Delta} = \underset{Iname}{O} = \underset{Takamači'a}{\Delta} = \underset{Ipiõčika}{O}$$

$$\Downarrow$$

$$\underset{Maihu}{O} = \underset{Čiami}{\Delta} = \underset{Iname}{O} \qquad \underset{Takamač\,i'a}{\Delta} = \underset{Ipiõčika}{O}$$

In this case, Iname's polyandry may be best interpreted as a gradual divorce from Takamači'a and a remarriage to Čiami. Since all of these individuals were part of a cooperative household, it may be that polyandry is a way to change marriage partners with less conflict, since ex-husbands and ex-wives may continue to live together. This is supported further by the history of the household. Čiami was previously married to a woman named Imuĩ. Čiami traded Imuĩ to Takamači'a in exchange for Maihu (Takamači'a's sister). Later, the brother and sister pair, Iname and Takwariči'a, entered the household. Takwariči'a gave his sister Iname to Takamači'a in exchange for Takamači'a's wife, Imuĩ.

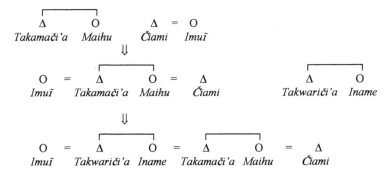

The Guajá may change marriage arrangements often, and this is not a complete history of the marriages of these individuals. For example, Maihu was previously married to an older man who died. One of Maihu's daughters was at one time married to Takamači'a in an avunculate marriage. Takamači'a's wife, Ipiõčika, is also a classificatory daughter of Maihu, also making it avunculate.

Several additional temporary polyandrous marriages were recorded during the research period on the Caru reserve, but there were no examples of polyandrous marriages where a single woman had two husbands without at least one of her husbands having another wife as well. However, it should be noted that frequent polyandry was reported in a neighboring reserve after an outbreak of influenza had killed approximately two-thirds of the Guajá. A demographic shift occurred, with almost twice as many men as women remaining (Gomes 1991, 1996). Gomes described the polyandrous unions there as creating a great deal of tension among the younger males who had difficulty obtaining wives. Although there were no examples of pure polyandry on the Caru Reserve during the research period, it is likely that this has been practiced before. When the husband of one young girl died, the girl's father lamented that his daughter's husband was dead and that she did not have another one. While the girl indeed had other classificatory husbands, she did not have another true husband, and the father's response suggests that polyandry is not necessarily considered unusual by the Guajá.

Authority and Control in Sex and Marriage

Rivière (1984), following Turner (1979), has argued that some lowland South American groups can be considered to have a political economy of women, where men control women as a valuable resource for their productive and reproductive abilities. Guajá men do exercise control over women through their role as the exchangers of women for wives. However, although men exchange women in marriage, they are not chattel in this operation, and women exercise control in a number of ways.

Women are exchanged in marriage by a brother, a father, or a husband. A young girl's first marriage is typically an agreement between the prospective husband and the girl's father. The father who gives the girl in marriage is the father who is responsible for raising her, whether or not he is a "biological" father in the Guajá sense of the word. Thus, it is the mother's husband, or the residential father who decides. When girls are older, they are traded by their brothers if their husbands die, and they can also be traded by their husbands.

Although men trade women as marriage partners and are able to exert control over the labor of their wives and daughters, they do not have the same degree of control over their sexual behavior. The Guajá have a separate term for their lovers, *mapara*, who may or not be their spouses. Due to Guajá beliefs about the necessity of multiple male sexual partners to bring a woman's pregnancy to term, it is expected that men and women will have several *mapara* sexual relationships outside of marriage.

The degree to which jealousies occurred over the romantic relationships of spouses with others was difficult to determine, because the Guajá were often reluctant to express negative emotions. One older husband did confide that he was extremely upset about the sexual relationship his twelve year old wife was developing with a fourteen year old boy. He dealt with this by leaving with his family to go on hunting trips whenever the boy was in the village. He felt powerless to control his wife's sexual behavior directly, but he was able to thwart the relationship by controlling his wife's labor and taking her with him on hunting trips and away from the boy. It was my impression that *mapara* relationships often lead to a marriage arrangement between the partners.

Even if jealousies occur, *mapara* relationships seem to be generally tolerated and accepted as normative behavior. Brothers freely told me that they were partial fathers of each other's children, women demonstrated no reluctance in naming the father sets of each of their children, and even the children were able to name the father sets of members of the group (for all, the names could only be given if the fathers were still living). In one household that included three *harapihára* brothers and four *harapihára* sisters, each of the sisters had sexual relations at one time or another with each of the brothers, while one of the brothers was serially married to three of the sisters. And, as mentioned previously, mothers and daughters living in the same household sometimes share sexual partners, and even the same husband. It should be noted that there was one individual who was reported to be extremely jealous of his wife's sexual partners. However, he displayed many antisocial behaviors, including habitual lying, and he had numerous conflicts with many of the other Guajá. According to informants, he brutally beat his first wife while she was pregnant, leading to the death of both the mother and the child. During the research period, he did the same to his second wife, leading to the death of the fetus and nearly to the death of the wife. FUNAI documents indicate that he, at one point, threatened officials with a knife. These behaviors were extremely atypical for the Guajá, and while jealousies certainly occur, such violent reactions to them seem to be rare.

It should also be noted that consanguineal links among the living were sometimes blurred. The father sets provided by one informant might differ from that of another. For example, one individual told me he was a partial father to two of a woman's daughters, but the woman told me that he had only fathered the younger one. Similar confusion of father sets occurred among other members of the community. "Confusion" is probably not the most appropriate term, rather, the ambiguity seems to be another aspect of the general flexibility in reckoning relations.

The avunculate marriage practice may foster extramarital relationships as well. Men who marry children may have to wait many years until their wives reach sexual maturity and therefore seek other sexual partners before that time. When the girls do reach sexual maturity, they may not be attracted to their older husbands, particularly when their older husbands have partially raised them and played a fatherly role. The *mapara* relationship, in contrast to marriage, tends to involve individuals who are of a similar age, although not always. While the romantic love of the *mapara* relationship has an important role in Guajá reproduction, the marriage relationship is the fundamental economic unit, with a wife's husband having the social responsibility for her children.[4]

The Guajá practice a form of brideservice which involves clear female control in the transition into marriage. Culturally, the Guajá describe the brideservice practice as a prospective husband bringing food to a young girl so that she will grow and not be afraid to go and live with him. For the prospective husband, he is demonstrating his ability to take responsibility for the child and to provide sufficiently for a wife. However, the young girl has considerable control in the process. While she is not able to choose her own first marriage partner, she controls the timing of when she is ready to go and live with her husband. As long as the child expresses fear, she may remain in the natal household. Theoretically, the girl could remain afraid indefinitely. However, usually after several months she starts to begin living, at least part-time, with her husband. Usually other female relatives are already living in the husband's household, which eases the transition. In addition, although young brides sleep at their husbands' houses, they often spend the day with or near their mothers, helping them in subsistence activities or caring for their younger siblings. Also, although girls have less freedom than do boys, child-wives also spend a good bit of their time just playing with other children.

The Guajá lack a formal marriage ceremony, which may have a relationship to their marriage practices in general. For example, the Trio, who also practice the avunculate marriage, have no formal marriage ceremony (Rivière 1969), nor do the Yanomamö, whose men also marry young girls whom they partially raise (Chagnon 1997). Among the Matis, child betrothals exist and young girls are raised by their fathers-in-law (Erikson, in press).[5] Taylor (2001:47) has described these child marriages among the Jivaro as "pygmalionism," in that it involves training young girls to be good wives. In general, among the Guajá marriage is more of a process than a strict rite of passage. Marriage as a process is reflected not only in the transitional marriage period of young girls, but also in temporary polyandry involved in the dissolution of some marriages and the underlying shifts of residence that accompany these transitions.

AFFINITY AND CONSANGUINITY

Viveiros de Castro (2001) has recently proposed what he calls the GUT ("Grand Unified Theory") of Amazonian sociality, which is extremely enlightening for understanding the Guajá kinship system. Here, he argues that affinity is the generic mode of relating in Amazonia. He presents a concentric model of Amazonian sociality whereby consanguinity itself is nested within affinity and is best understood as a special condition of nonaffinity. Further, consanguinity and affinity should not be considered as discrete categories; rather they lie on a continuum and are in perpetual disequilibrium. He argues that there is no relation without differentiation. Taylor (2001) echoes this in describing pure consanguinity among the Jivaro as a nonrelation, analogous to cloning.

Viveiros de Castro (2001) contrasted three types of affinity in Amazonian sociality: potential affinity, virtual affinity, and actual affinity. Originally, potential affinity referred to a generic mode of affinal relatedness which he contrasted with virtual affinity, the terminologically encoded affines, and actual affines, those who were directly linked by marriage, such as the brother-in-law. As he notes, the terms potential and virtual have sometimes been used interchangeably in the literature, and he suggested that meta-affinity may be a more apt term for potential affinity and that he now prefers to use virtual affinity to refer to the precosmological background from which affinity arises.

Following Viveiros de Castro, I am distinguishing three types of affinal relations among the Guajá: meta-affinity, terminological affinity, and actual affinity. I am taking some liberty with the terms, primarily to avoid confusion between the way the terms potential and virtual affinity have been used. In doing so, I hope that I am not distorting the meaning.

Meta-affinity refers to Viveiros de Castro's concept as affinity as a generic mode of relating in Amazonian groups. It is the fundamental form of relating which informs all others. The concept is apt for the Guajá in their general application of affinal terminology to nonhuman plants, animals, and supernatural beings. This will be taken up further in chapter five.

The distinction between terminological and actual affines exists within Guajá terms of reference and address, with actual spouses receiving a different term from those who are possible or former spouses. However, these distinctions are also malleable due to plural paternity. A sister's daughter is marriageable for a male ego, but for the Guajá, roughly two times out of three, the father of the child is not her actual husband nor her brother's actual brother-in-law. A sister's actual husband can be a partial consanguine to a male ego if the two men share a father.

Here, Viveiros de Castro's description of affinity and consanguinity as lying on a continuum is apt for the Guajá. Plural paternity creates multiple pathways of relating to any given individual. Alternate paternal pathways are often used to classify a particular individual as a possible sexual or marital partner. For example, one woman described her relationship differently to two brothers who shared the same mother and one of their fathers and who were raised by that same mother and father. The father described of the brothers was also a classificatory father to the woman. She traced her link to the younger brother through one of his other "biological" fathers in another village and called him "husband." However, she traced her link to the older brother through her classificatory father and called him "brother." She could have just as easily justified calling the younger brother "brother" by tracing the link through her classificatory father and called the older brother "husband" by tracing the link through another "biological" father.

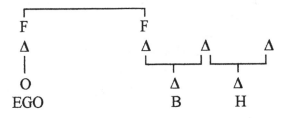

Contrasting this background of meta-affinity as a generic mode of relating is the often-observed tendency for Amazonian peoples to use consanguineal terms for real affines with which they live (e.g. Conklin 2001; Taylor 1998; Viveiros de Castro 2001). Conklin (2001) describes the consanguinealization of affines among the Wari' as a biosocial process. Here, it is not a social fiction but is viewed as a ongoing biological process that occurs through the sharing of bodily substance with affines (as well as consanguines) living together. Husbands and wives become consanguinealized through the exchange of bodily fluids in sexual intercourse; mothers and their children become consanguinealized through breast milk; fathers can pass invisible blood elements to their children during sleep; and, more generally, even sweat can serve to transmit bodily substance between individuals.

One area where consanguinity is important is in acquiring access to foraging areas. Gomes (1996) observed that despite the nomadism of the Guajá, they do have a notion of personal rights to particular areas of the forest. I am somewhat hesitant to use the term territory because the term suggests fixed territory, territoriality, and corporate kin-group ownership of territory, none of which apply here. In addition, while consanguinity is important in transmitting rights to a given territory, affinity also plays a significant role. Further, there is also an achieved, individualistic aspect attached to the foraging area that individuals exploit, which creates flexibility in its boundaries. In this light, its egocentrism and flexibility make it not unlike Guajá kinship itself.

Guajá use of "territory" is also consistent with the notion of the common property regime, which is often found in hunter-gatherer groups (Lee and Daly 1999). According to Lee and Daly, a prevalent pattern among hunter-gatherers cross-culturally is to hold land in common by a kin group rather than individuals having unlimited access to communal land. However, this is neither primitive communism nor is it corporate ownership of territory. While there may be a de jure claim to land by a kin group, it is de facto communal property due to demand sharing, which is also characteristic of hunter-gatherers. Ingold (1999) has described demand sharing as the tendency of hunter gatherers to share through mutual taking rather than mutual giving. Thus, all one would need to do is ask permission to use the land of another; there are no real owners.

The Guajá refer to their personal foraging areas as hakwa or harəkwa, which translates as "mine known" or "same-sexed sibling known." In addition, they call the land of the Guajá people as a whole Mai'irakwa (known by the creator divinity). Individual foraging areas are not named. Although the Guajá are generally peaceful, they do defend what they perceive to be the boundaries of their land. In one FUNAI report of first contact with one Guajá family, the tools that were given to them were used for a murder relating to land boundaries

(Damascena da Silva 1989). Several informants also told me of the incident and described the individual who died as the former husband of two women living on the Caru Reserve. In order to introduce themselves to the group, the FUNAI officials presented a family with gifts, including an ax, and, subsequently, the ax was taken and thrown at another Guajá getting honey from a tree within the boundaries of their foraging area, killing him. The Guajá will also defend the reserve against non-Indian Brazilians. During the research period, three Guajá men murdered a Brazilian man who was fishing on the reserve, which the Guajá now perceive as the boundaries of their land.

The Guajá practices differ from a strict definition of the common property regime because they lack well-delineated kin groups. However, foraging grounds are initially transmitted through consanguineal relationships. Sons and daughters have the right to forage in the areas of their "biological fathers." It would probably be a distortion to describe this as "inheritance," but it is clear that access to foraging areas is not a random choice but is mediated by consanguineal ties.

When children marry, actual affinal ties provide them with additional options for foraging grounds, for husbands and wives have the right to exploit the foraging grounds of their spouses. Thus an individual with two fathers who marries an individual with two fathers would potentially gain access to four foraging areas. Further, the avunculate marriage pattern often provides individuals with access to at least some of the foraging areas of their mothers. If a female ego marries her mother's brother, she gains access to the foraging area of her mother through her spouse. Brothers and sisters often trek cooperatively, and, for a male ego, the marriage of the sister to the mother's brother gives access to his mother's foraging area.

Although access to foraging areas is initially determined by one's patrilineal links, spousal rights, particularly through the avunculate, create a system that is bilateral in function, although not in structure. In practice, individuals take advantage of the realm of kinship relations to gain access to resources and to disperse themselves across the landscape. The pattern is somewhat similar to Ju/'hoansi (!Kung San) flexibility in residence patterns by using kinship affiliations to disperse groups among waterholes and other resources (Lee 1968).

Another important aspect of the *hakwa* is the degree to which it is understood as achieved. A *hakwa* cannot be possessed by children or young adolescents because they are said to be too young to have had sufficient experience to be familiar with the forest. Despite the paternal transmission, it is conceptualized much more as "mine known" than "mine owned." The experiential aspect of the *hakwa* also provides more fluidity in the boundaries of a foraging area, allowing for the possibility of range expansion and migration. Thus, the system does not resemble estate inheritance in Western society, where land, with clear boundaries, is parceled off to descendants. Acquiring a *hakwa* involves a combination of experience with an area of forest within one's own lifetime as well as negotiation of social relationships for access to foraging areas in order to disperse individuals across the landscape. Informants described such negotiation taking place when some Guajá were transferred from other areas to the reserve. In addition, although the Guajá state that they must ask permission before foraging in another's *hakwa*, given the cultural values on sharing, as well as the flexibility in their kinship links, most Guajá probably have a means of access to most of the foraging areas of other Guajá. However, considering the ax murder, some limitations clearly exist.

The question does arise as to whether the notion of the *hakwa* can be reconciled with the traditional nomadism and known migrations of the Guajá. The *hakwa* does not suggest that foraging areas are immutable; rather, it suggests that foraging areas are mediated through social relations and are not merely chosen at random. Further, since actual experience with an area of the forest is a prerequisite for claiming a *hakwa*, migration and subsequent knowledge of a new foraging area certainly could take place.

Consanguinity also plays a role in men's sharing of fish and game meat, which follows a somewhat similar pattern to the way access is obtained to foraging areas. While it is obligatory to share food if someone asks, a pattern does exist for who one first asks. Beyond the household, a father's consanguineal children are often the first to arrive to ask for a portion of his catch. In many cases, a consanguineal child will arrive with his or her partial consanguine who is not a consanguine of the father. So a consanguine of a consanguine is a kind of consanguine by proxy.

GENEALOGICAL AMNESIA

Reconstructing Guajá genealogies was difficult not only due to the multiple pathways created by plural paternity, but also due to their profound genealogical amnesia (see also Cormier 1999). "Genealogical amnesia" is the term often used to describe the markedly shallow depth of recall of forebears, and it is a widespread characteristic among cultures of lowland South America. The term was coined by Geertz and Geertz (1964) in describing the shallow depth of recall of kin among the Balinese. Geertz and Geertz drew on Barnes's (1947) description of what he called "structural amnesia." It is interesting to note that Barnes's original use of the term was in describing unilineal descent groups who were able to trace consanguineal vertical links back multiple generations while truncating affinal lines. In this respect, his structural amnesia refers not so much to the ability to recall, but to the social significance of recalling or not recalling ancestors in creating certain types of kinship systems.

The Guajá's genealogical amnesia is extreme. Typically, when I asked someone the name of a deceased relative, the answer was to gesture upward and say, "*nɔkwai, oho iwabe*," meaning that the relative was in the celestial sky home, which was considered sufficient explanation. The Guajá have no taboos on the names of the dead. They would make an effort to remember the names when pressed, but most had difficulties recalling names. No one was able to recall the names of members of the second ascending generation, and most had difficulty recalling the names in the first ascending generation, even those of their own parents if they had been dead for more than a few years. I was able to construct some genealogical history through using FUNAI records and comparing the partial recollections of multiple informants. If I had a name, particularly the name of a parent, the Guajá were usually able to confirm that individual as their parent. In addition, the Guajá would sometimes come back later and give me a name after they had consulted with each other. But clearly, their recall of names was an attempt to be helpful to me. Genealogies are neither meaningful or appropriate for understanding the way the Guajá perceive kinship relations. The relevance of kinship ties among the Guajá must be understood in terms of what information is available about the living members of the group.

Employing Barnes's framework of structural amnesia requires that one understand the cultural uses of remembering certain genealogical links. For example, Sahlins's (1961) classic work on the patrilineal Nuer and Tallensi described the segmentary lineage as a means of sociopolitical organization and territorial expansion. For the Guajá, the inability to recall one's ancestors may as valuable an organizational principle as the ability to recall is in others. Specifically, one of the important consequences of genealogical amnesia is that it allows more flexibility in reckoning kinship relations for sexual and marital partners.

Although genealogical amnesia is widespread in lowland South America, the cultural practices that foster it differ. Some cultures have outright taboos on the names of the dead, such as the Mundurucu (Murphy 1956) and the and the Yanomamö (Chagnon 1997). Another common practice is teknonymy, whereby parents are named after their children. Teknonymy is associated with genealogical amnesia and seen in a number of lowland South American groups—the Araweté (Viveiros de Castro 1992), the Ka'apor (Balée 1994a), the Mehinacu (Gregor 1974), the Mundurucu (Murphy 1956), the Sirionó (Holmberg 1969), the Wari' (for mothers only) (Conklin 2001), and the Yanomamö

(Chagnon 1997)—and in other parts of the world, such as Bali (Geertz and Geertz 1964), Borneo (Needham 1954), and China (e.g., Feng 1936).

Viveiros de Castro (2001) makes the point that teknonymy also serves to con-sanguinealize affines. One might even consider it a reversal of descent, in that identity is constructed through one's descendants rather than through one's an-cestors. Teknonymy restricts the retention of genealogical information because the names of forebears are merged with those of the living, particularly when original, given names fall out of use. In Bali, Geertz and Geertz (1964) specifically argue that teknonymy acts in conjunction with genealogical amnesia to allow for flexibility of membership in Balinese corporate kinship groups. Balée (1994a) has argued that teknonymy reinforces the marriage bond among Ka'apor parents, which also suggests that it has an important role in alliance formation.

Other naming practices that serve to foster genealogical amnesia, and poten-tial reclassification of a kin relation, include inherited names and multiple name sets. For example, the Mehinacu of the Upper Xingú inherit names from the grandparents or other collateral relatives in addition to practicing teknonymy (Gregor 1974). The Suyá of Mato Grosso have inherited name sets that may involve up to forty-four names for a single individual (Seeger 1981). Seeger also describes frequent reclassification of relatives by the Suyá. Outside of the natal household, the Suyá reclassify any sexual partner as a "spouse," in-cluding the female patrilateral cross-cousin, who in their terminological system is a classificatory mother for a male ego. When the same name applies to more than one individual (inherited names) or when single individuals have multiple names (name sets), it is difficult to recall a specific individual by referring to his or her name alone.

SIBLINGSHIP AND THE RESIDENTIAL GROUP

The core of the Guajá's residential unit is formed by one or more nuclear fami-lies that are typically connected through siblings. The siblings may be a brother and a sister or may involve two or more brothers or two or more sisters. Perhaps most broadly, residence can be described as neolocal, as new couples have a household separate from those of their respective parents; however, other labels can also be attached to their residence patterns.

A pattern resembling natolocal residence occurs early in the marriage. After a marriage is arranged, the husband and the wife initially stay in their respective households. No clear-cut line exists between when a betrothal ends and a mar-riage begins. Natolocal residence occurs primarily before the couple is considered truly married, but this varies with informants and with the particular couple at hand. Further, early in the marriage, a young girl may alternate back and forth between staying in her parents' home and her husband's home. In addition, a young married girl may spend the day working alongside her mother, or both parents, and then sleep at her husband's home. Or, she may spend only part of the day with her husband, and sleep at her parents' house.

The residence pattern largely depends on the age and marital status of the individuals in question. A young man who is seeking his first child bride may have an element of uxorilocal residence associated with his brideservice. A prospective groom may hunt and fish with his father-in-law and sleep some nights at the home of his prospective wife's parents. On several occasions, men were noted to stay a week or so at the home of their prospective brides during this transitional time before returning to their own households. An older, more established man, with an older wife and his own household, may perform no brideservice at all and may acquire a child wife through trading a daughter, sister, or wife.

When a young girl moves into the household of an older man, she is often moving into a household that already contains many of her own female relatives, another element of uxorilocality. Typically these are her sisters, but her grandmother may also be in the household. The same may be true when a girl's husband dies. Due to the age discrepancy at first marriage, girls typically outlive their first husbands. A widowed woman will often move or marry into a household with female relatives, such as her mother, daughter, or sister. In addition, households may have related brothers. Often, related brothers will marry women who are related as sisters, since a set of brothers may stand in a similar marriageable relationship to a set of sisters, although this is often complicated by the kinship networks created by plural paternity.

One example of a household built around related women involved a grandmother, the grandmother's daughter, and the grandmother's two granddaughters. The grandmother's daughter was the father's sister to the two granddaughters. The grandmother and one of her granddaughters were married to the same man. The other granddaughter was married to the same husband as the grandmother's daughter; that is, she was married to the same man as her father's sister. The grandmother's daughter had several children partially fathered by the grandmother's husband. The grandmother's husband was also a partial father to the granddaughter he was not married to. Here, the granddaughters were two paternal sisters who were only a year apart in age and grew up together in the same natal household with the same mother and her husband, who was one of the fathers of both the girls. The elder sister married her grandmother's husband, which suggests uxorilocal residence. The younger sister, although she also moved into the same household, could be considered more virilocal, since she moved into the household of one of her father's sisters and married her father's sister's husband. Some of the children of her father's sister were also her true brothers since they shared a father. In addition, there was a betrothal of the grandmother's daughter (the father's sister of the granddaughters) of her own granddaughter to her second husband, creating another grandmother-granddaughter union.

Perhaps what is more important than attempting to classify the residence patterns is understanding that the Guajá residential unit, regardless of type, overrides genealogical relationships to some extent. The term of address is more strongly related to residence than it is to genealogy. As with many Cariban groups, therefore, coresidence creates fictive kin ties (e.g., Basso 1973). Viveiros de Castro (2001) refers to this process as consanguinealization. Generally, same-sexed individuals in the same household, who are either of the same age or who fill a similar social role, refer to each other with the same-sexed sibling term of address, *harə*. While the term *harə* properly refers to a maternal or paternal same-sexed sibling, it is used to refer to anyone who in some way acts like a same-sexed sibling. Thus, brothers-in-law who coreside and are in a cooperative relationship with each other may call each other *harə*, rather than the typical brother-in-law term of address, *čiá*. Another example involves the previously mentioned grandmother and granddaughter pair who were married to the same man simultaneously. They referred to each other as *harə* rather than using grandmother and granddaughter terms of address, because their relationship to each other as co-wives had an equivalence similar to that of sisters.

Changes in the term of address between two individuals due to marriage ties and coresidence is a clear source of ambiguity. The social relationship established through the residence pattern may take precedence as the actual genealogical links become less important or are forgotten. Erikson (in press) described a similar phenomenon among the Panoan-speaking Matis. Here, when conflicts occur between genealogical and social ties, the social ties take precedence. This is also similar to Rivière's (1969:65) description of the Trio: "Coresidence can be as binding as the ties of genealogical connexion and, in Trio thought they are not truly distinguished."

TERMS OF REFERENCE AND ADDRESS

In addition to changes in terms of address in the residential group, terms of reference and terms of address often do not always coincide exactly, leading to more ambiguity. Broadly, terms of reference tend to describe a specific consanguineal or affinal relationship while terms of address mark how one should behave towards another.

The parental and grandparental terms of address suggest the level of respect one should pay to a given individual, rather than specific consanguineal or affinal information. The two basic address terms for parents are *čipa* (father) and *amã* (mother). The three forms of the *čipa* term indicate the degree of respect given: *čipa'i* (father [diminished]), *čipa* (father [unmarked]), and *čipa-te* (father [emphasized]). The suffix -*te*, also carries the meaning of "true" or "truly," with the suffix -*'i* contrasting as "somewhat." The term of address *čipa-te*, (father [true]), differs from the term of reference *tu-te* (a genitor) and is contrasted with *tu-na* (a classificatory father [FB, MH]). The suffix -*na* negates and can be glossed as "false." Likewise, the term of reference *ihɨ-te*, is used for the birth mother while *ihɨ-na* is used for classificatory mother (MZ or FZD).

The *čipa* terms do not refer to specific genealogical relationships, but to the degree of deference one should show to the individual. Typically, one uses *čipa'i* for a husband, *čipa* for a father, and *čipa-te* or *čipa-tamə* for a grandfather. However, there is considerable variation within the categories. What is most striking is the use of a father term for the husband. Such a designation reflects, in part, the paternal role of the husband in relation to a young girl who first marries at around six or seven years of age, often to a considerably older man. Husbands of young girls behave more as fathers towards spouses until they reach sexual maturity. Several groups have similar practices wherein older men marrying younger girls that they partially raise, including the Yanomamö (Chagnon 1997); the Tapirapé, who call it "raising your own wife" (Wagley 1983:157); and the Trio, where it is specifically the sister's daughter who is raised (Rivière 1969:161). The use of the father term by young girls toward their husbands describes the age discrepancy and the fatherlike role played by the much older husband. However, all women use a father term for their husbands, even when remarried to a husband that is significantly younger than they are; so the use of the term encodes the meaning of a man toward whom the wife should behave respectfully.

The *čipa'i* term, then, refers to one to whom one gives respect, but to a somewhat lesser degree than certain other males to whom one gives respect. While *čipa'i* is most commonly used by a woman to refer to her husband, the status of the husband within the group as a whole, relative to her status, also affects the term of address used. For example, one young woman married a man who was considerably older and was in the *tamə* category of older men. She called her husband *čipa-te* rather than *čipa'i*. However, the man was also married to another woman, closer to his age, who was in the *iari* category of older women, and she called her husband *čipa'i*. In addition, the children (wife's children and his own) of this man often called him *čipa-te*, instead of the typical unmarked *čipa* for father, which indicated the man's status in the group. Fathers who are very young (about seventeen to twenty-one) are usually called *čipa'i* by their children, which suggests a diminishment of their relative "*čipa*-ness" due to their youth. In addition, *čipa'i* may be used instead of *čipa* for a classificatory father, which, rather than indicating "false-father," as with the *tu-na* classificatory term of reference, indicates a father to whom ego acts with somewhat less respect than a "biological" father. One further important point is that only the men to whom women are actually married receive the *čipa'i* term. Terminological spouses with the *iménə* term of reference do not have a general term of address and are called by their personal names.

The *amã* terms of address that show respect to women do not correspond exactly to the male *čipa* forms. The majority of women who receive this term receive the unmarked *amã*, while only very elderly women receive the *amã-čiá-te* term, and no diminutive form was observed (that is, no *amã'i*). An individual will call his mother and his mother's sister *amã*, but others who fall into the *amã* category are less clear. Largely, this term of respect is given to women who are at least of childbearing age and are also older than ego. Males, and particularly very young males, may refer to their father's sisters as *amã*, although many also use the sister term, *čikari*. The term for father's sister's daughter, who receives the same term of reference as the mother, *ihɨ*, also varies. Women will sometimes address this individual with the sister term, *harə*, if the two are close in age, which is often the case for cross-cousins. Men may also address this individual with the sibling *čikari* term. One informant described this individual as *amã-čikari* (mother-sister). Further, since it is possible for a man to marry his patrilateral cross-cousin (who, like a mother, is an affine), her term of address would then change to the spousal term.

The sibling terms of address, while indicating equality for same-sexed siblings, also evoke an element of inequality among opposite-sexed siblings. Likely, this is related to the brother's ability, like a father, to trade a sister in marriage. The same term, *čikari*, is used by brothers for their sisters and by fathers for their daughters. Mothers also call their daughters *čikari*, which indicates some degree of maternal authority over daughters. In addition, all male children tend to call all female children of about their same age *čikari*, and the girls call the boys by the brother term, *či*. Although even young children can describe who would be in the spousal category, opposite-sexed young children have sibling-like relationships with each other and use sibling terms of address rather than spousal terms for each other. By the same token, a man may refer to a small female child in the spousal category as *čikari*; the term of reference, *emeriko mičika'ɨ* (little wife) is also used. Further, there is a general term of address used by anyone to refer to a young child, particularly toddlers: *hača'a*. Parents, older siblings, and distantly related individuals all call small children by this term.

As described previously, the *harə* term of address refers to someone who is in a relationship to ego like that of a same-sexed sibling. Both men and women use the term to refer to their paternal same-sexed siblings, their maternal same-sexed siblings, or to same-sexed individuals of about the same age with whom they cooperate, if they do not fall into the *harəwaiya* category of MB/ZS for a male ego or FZ/BD for female ego. However, coresidential *harəwaiya* are often called *harə* rather than *čiá*, particularly males who live with their female siblings. For example, in one household, there were five men and their wives. Four of the wives were paternally related siblings, *harəpihárə*, and were fairly similar in age, while the other was a young girl of about seven who was more distantly related and in transition to becoming the wife of the youngest of the males. Three of the men in the house were paternal siblings, while the other two are not closely related to either the brothers or to each other. However, all the men refer to each other as *harə*. When asked, they will distinguish between *harə* and *harə-na* (false *harə*), but in common speech there is no such distinction.

Another important point is that while women use the *harə* term to refer to someone who either is or behaves like a same-sexed sibling, males have an additional term of address that distinguishes the maternal same-sexed sibling, *atɨ*, which is the same as the term used for a son. However, the maternal same-sexed sibling may also be referred to as *harə*. Thus, the term *atɨ* can be used for both a consanguine or an affine. The term *čikari* is similar in that males use it to refer to consanguines (sisters and daughters) while females use it to refer to their daughters, who are affines. However, when *atɨ* and *čikari* are used for affines, they are those affines with whom one may have a particularly close relationship, such as mothers and their own daughters and their sisters' daughters. Maternal same-sexed siblings often grow up in the same household and definitely grow up in the same household if they share the same mother, regardless of whether the fathers are different.

Terms of address for both men and women define their behavior towards those in the spousal category. For a man, the term of reference *emeriko* defines an individual as a possible marriage or sexual partner, but the terms of address *čiku* and *čia'ə* describe how he is to behave towards her. The *čia'ə* is used for the true wife, a woman with whom a man enters into a cooperative social and economic relationship. This is the woman he lives with, he treks with, and whose children he cares for. The *čiku* term describes a woman as a real or possible sexual partner who is terminologically a spouse, but to whom he is not currently married. As previously mentioned, women also make a distinction in their terms of address between a man they are married to and those that are sexual partners, possible sexual partners, or possible marriage partners.

AMBIGUITY

Ambiguity is a key characteristic of Guajá kinship due to dual classification through the avunculate marriage, plural paternity, incongruities in terms of reference and address, profound genealogical amnesia, and the residential unit's overriding genealogical ties. In general, kinship relations are flexible rather than absolute and are individualistically determined, with even same-sexed siblings rarely sharing the same kin universe. Outside of Amazonia, the Mardu Western Desert Aborigines have a structurally similar egocentric construction (Tonkinson 1991); however, their system is not due to multiple genitors. Tonkinson (1991) describes the Mardu as bilateral, but their system exhibits unilineal aspects in that they have a choice as to whether they reckon kin relations through their mothers or their fathers. Like the Guajá system, Mardu kinship is extremely flexible, and same-sexed siblings do not necessarily share the same kinship universe (Tonkinson 1991:52). While ambiguity among the Guajá seems to be primarily a means of manipulating consanguineal and affinal relations for marital and sexual partners, it also functions as means of social control and maintaining the integrity of the residential unit.

The Guajá reclassify some individuals to increase social distance as a means of social control. The Guajá, as a traditional foraging people, lack any true authority or leadership position. Although the opinions of elder men and women are respected, they have no formal authority, nor is there a formal mechanism for punishing violators of norms. Generally the Guajá make efforts to avoid conflict with each other. It was rare to see overt disagreements among individuals or even the expression of negative emotions. A similar reluctance to express strong negative emotions was also noted among the linguistically closely related Kagwahiv (Parintintin) (Kracke 1978). Teasing and joking, however, were quite common among the Guajá and seem to be an important mechanism for decreasing social tension.

The Guajá also use the flexibility in reckoning kinship relations to decrease social contact and social obligations with those who violate cultural norms, thereby decreasing social tensions. For example, an individual was described previously who displayed antisocial behaviors and had committed murder and attempted murder within the community. When collecting kinship information, many Guajá either claimed not to have a kinship tie with this individual or they stated they did not know how they were related to him, although they had no problem in stating their relationship to the individual's close relatives, such as his brother, wife, and children. Further, the individual could name kinship ties to most of the Guajá who were claiming not to have or know a relationship to him. While there was no outright ostracism of the man, alternative classification allowed many of the Guajá to sever their kinship links to him and defuse social tensions by increasing social distance. Ambiguity, then, allows kinship classification to be used as a negative sanction for unwanted behaviors.

Ambiguity in kinship classification may also facilitate maintaining the integrity of the residential unit. Previously, it was stated that residence overrides descent. Perhaps more accurately, residential relationships are modeled on kinship relations, and kinship terminology provides the foundation for establishing appropriate behaviors among co-residents. Coresidential women are *harə* to each other as are coresidential men. Coresidential men and women relate to each other as brothers and sisters and as husbands and wives. The sibling relationship is as important as the marriage relationship. While the marriage relationship provides an economic division of labor, real or fictive same-sexed siblingship provides a cooperative partner in labor. Further, the opposite-sexed sibling relationship is key in reproducing marriages as the sister, through her daughter, can provide a wife for her brother. The avunculate marriage, then, strengthens the brother-sister bond as well as the bond a man has with his sister's husband, who may well be his mother's brother.

Although the avunculate pattern represents a form of matrilateral cross-cousin marriage since ZD is simultaneously MBD, it does not represent generalized exchange (Lévi-Strauss 1969) for several reasons. First, generalized exchange suggests exchange between descent groups, which the Guajá do not have. Second, functionally, the system operates much more like direct exchange, in that where a male ego gives a sister, he receives a wife. The latter point was noted early on by Lévi-Strauss who interpreted the avunculate marriage as a form of the patrilateral cross-cousin marriage. He described it as the claim of a man who gives his sister and expects his sister's daughter in return for either himself or for his son (Lévi-Strauss 1969:447). Claiming the daughter for himself was seen by Lévi-Strauss as a privilege allowed in the avunculate, which stems from emotional needs for immediate gratification. According to him, the avunculate marriage is

> at once a greedy and individualistic attitude . . . the giver seeks compensation immediately, or as quickly as possible in a form which shall maintain in the highest degree the concrete and substantial tie between what has been given and what should be returned.
>
> (Lévi-Strauss 1969:448)

A similar interpretation was made by Laraia (1971:197–98) in studying Tupí kinship, finding that delayed exchange (FZD) marriage could be seen as a man giving his sister to a lineage and, in return, getting a daughter from the other lineage (that is, a female of the next descending generation). Following Lévi-

Strauss, Laraia viewed this in terms of a selfish father, who here took the daughter meant for his son because of a desire for polygyny.

While the ZD marriage does seem to function as a system of direct exchange rather than generalized exchange, an interpretation in terms of immediate gratification for a male ego does not seem applicable to the Guajá. As previously mentioned, a young girl's first husband takes on more of a paternal role than a spousal role until the girl reaches sexual maturity, which could take six years or more. Rather than immediate gratification this could even be interpreted as a form of extended brideservice. Lévi-Strauss's interpretation of the avunculate as "maintaining in the highest degree the substantial tie between what has been given and what should be returned" (1969:448) is key in understanding the Guajá avunculate. However, the tie is not between kin groups, nor perhaps so much between the men involved in the exchange; cementing the bond between brother and sister seems just as important.

5

ANIMISM AND THE FOREST SIBLINGS

The relationship of Amazonian peoples to local plant and animal species has become a subject of increasing interest among Amazonian researchers. In ethnobiology, both within and outside of Amazonia, researchers are demonstrating an increasing awareness that preservation of biological diversity is inextricably linked to the preservation of human cultural diversity. While this might seem self-evident, it is complicated by the attributing a Western conservation ethic to indigenous peoples. Redford (1991) described this in his article on the "ecologically noble savage," and Descola (1998) has addressed similar themes with special attention to Amazonia. Thus, it is of paramount importance to understand the perception of the environment from the indigenous point of view. A fundamental difference in Amazonian thought is that categories that are often taken as structuralist givens such as the nature/culture boundary, the affinity/consanguinity boundary, and the self/other boundary are perceived more as a continuum than the Western either/or distinctions allow for.

ANIMISM

One of the most important recent contributions to understanding the relationship of Amazonian peoples to the environment has been the reformulation of

FIGURE 5.1   The squirrel monkey (Samiri sciureus)

animism, particularly in the work of Descola (1992, 1996, 1998), Viveiros de Castro (1999a, 1999b), and, outside of Amazonia, Bird-David (1999).[1] A central theme in their work is that animism does not merely consist of attributing a spiritual nature or other human feature to nonhumans in the abstract, but is fundamentally about social engagement. This reformulation of animism is beginning to shed new light on the prey-pet paradox in Amazonia, cannibalistic cosmologies, and kinship construction.

Descola's Animic and Totemic Systems

Descola (1992, 1996) describes animism as a mode of identification that is a symmetrical inversion of totemism. He argues that while totemic systems model social relations based on the discontinuities in nature, in animic systems social relations within the group are extended outward to nature. While making it clear that considerable variability exists among Amazonian groups, he has classified three typical modes of relating to animals and, to some extent, plants: reciprocity, predation, and protection.

Descola (1996, 1998) classifies the Tukanoans as representative of the reciprocal mode in terms of their view of humans and animals as being equivalent units of energy exchanged in a closed ecosystem. Århem (1996) describes a reincarnation cycle of human beings and animals among the Eastern Tukanoan Makuna. When a Makuna dies, he or she is cannibalized by the gods, and the spirit is reincarnated into a new person. The Makuna perform the same function for their prey; consuming the prey allows for its reproduction through reincarnation.

The homeostatic ecosystem is also reminiscent of the Tsembaga Maring of New Guinea, who demonstrate reciprocity in both the cognized and operational domains, to use Rappaport's (1979) terminology. In the etic, operational model, Rappaport describes a cyclical ritual mode of production involving the Tsembaga Maring and their regulation of the domesticated pig population, warfare, and distribution of territory. In the emic, cognized model, the Tsembaga Maring are involved in reciprocal exchange with their ancestor spirits, where the ancestor spirits who assist in warfare and enhance fertility are repaid with the periodic slaughter of pigs following warfare, which the ancestors symbolically eat.

Descola (1996, 1998) describes the predatory mode as applying to warlike societies with a cannibalistic social philosophy, using the Jivaro as an example. He describes the Jivaro's extension of kinship relations to nonhumans, whereby domesticated plants are considered affines and forest animals are considered consanguines. According to Descola, the Jivaro do not employ the idea of reciprocal exchange either naturally or supernaturally in their ideas about killing.

Descola (1996, 1998) describes protectionism, or gift-giving, as a combination of predation and reciprocity, wherein nonhuman groups are seen as dependent on humans for their reproduction and welfare. He describes the domestication of plants and animals as the most clear-cut examples of animic protectionism, but also includes pets, sacred animals, and ancestral figures. He sees this protection as reciprocal to some degree, in that it provides a benefit to the protector. Further, he describes the protection given by humans to nonhumans as a reflection of the relationship of the deities to humans.

Tupinambá cannibalism and its relationship to the avunculate marriage, according to Viveiros de Castro's (1992) analysis, can likely be interpreted as a form of protectionism. Enemy captives were called brothers-in law and made into "pets" for the sisters of the captors. Prior to being eaten, the captive would father the children (sisters' daughters) that the captor would marry. Being cannibalized was considered a heroic death and also prevented the dead body from being eaten by malevolent spirits, which was the fate of those who died at home from illness or accident.

Tupinambá cannibalism incorporates both predation and reciprocity, which is characteristic of Descola's protectionist mode. Although we do not know if the Tupinambá applied kinship terms to nonhuman forms of life, it is interesting that the term "pet" is used for the captive, which could be as much a metaphorical extension from nonhuman affine as it is from the nurtured role of the captive. The captors benefited by providing women with husbands and men with sisters' daughters to marry. The captive is temporarily nurtured, but the more important perceived benefit is his being granted an honorable death and avoiding being eaten by malevolent spirits. The food value of the victim would seem here to be minimal, particularly given the costs of keeping a "pet" enemy until he produces a daughter. It is the victim who benefits by being eaten, and the captor provides a service to the victim, which can be considered a form of delayed reciprocity, in that the captor would hope that he would be captured later by the victim's group to die with honor and avoid having his own body eaten by malevolent spirits.

Of Descola's three animic systems, the Guajá conform best to the protectionist/gift-giving mode in their relationships to both the divinities and their pets. All of the *iwa* divinities are seen as benevolent and are petitioned for assistance in bringing game animals to them, in healing, and in arranging marriages. A similar relationship is found in the Guajá pets, who are nurtured as children. The Guajá benefit from nurturing the animals in that they provide companionship, function in the socialization of children, and are a symbolic adjunct to female fertility. The pets' benefit is not being eaten. Game animals can also be considered to benefit in that their consumption affects their transformation into a sacred form in the *iwa*.

Viveiros de Castro's Perspectival Multinaturalism

Viveiros de Castro (1998a, 1999b) explains animism in Amazonia in terms of what he calls "perspectival multinaturalism." The basic idea is that in Amazonia, all beings are viewed as having a similar, human-like spiritual component, consciousness, intentionality, and culture. Viveiros de Castro points to a common theme in Amazonian creation myths wherein animals are created from human beings. Thus nonhumans can be considered to be ex-humans or "persons." However, due to the transformation of some people into nonhumans, they differ in bodily form and therefore differ in their modes of engagement in the natural world.

Human and nonhuman beings do not so much differ in the essential values and categories that they impose on reality, but in their subjective points of view of the same reality. The natural world is a constant, the essentially human "nature" of nonhuman beings is a constant, but what differs is the subjective perspectives of nonhuman beings, experienced through their habitual modes of engaging in the world. Thus, it plays out for example, that jaguars see blood as manioc beer, tapirs see watering holes as hammocks, and game animals see themselves as human and may see humans as prey.

Unfortunately, I did not ask the types of questions during my field work that would allow me to gain an appreciation of how perspectival multinaturalism might inform Guajá understandings of monkey socioecology. However, in retrospect, a number of Guajá descriptions of other beings illustrate the principal. Generally, the Guajá view plants and animals as having the same type of social organization as they have. For example, neighboring babassu palms may be described as wife and husband to each other and two pet howlers may be described as *harapihára* siblings to each other. In addition, the perspectivism can be seen in Guajá descriptions of the spiritual journeys of underground beings. The Guajá themselves make spiritual journeys to their *iwa* sky home. The underground beings make a similar journey from beneath the earth to above the earth. What appears as the earth to the Guajá appears as the *iwa* to the underground beings. Perspectivism may also apply to Guajá view of the activities of their spirit doubles (discussed further below). The same activities that the Guajá engage in on earth are simultaneously performed by their present-time spirit doubles in the *iwa*. If a person sleeps in a Brazilian cloth hammock on earth, the person's spirit double is sleeping in a traditional *tikwita* hammock at the same time in the sky home. In addition, Guajá creation mythology (see chapter seven) does hold that many of their prey animals originated from humans. Howlers are referred to as *awa-kera* (literally, "formerly human"). Most other monkeys and prey animals have undergone two transformations, from human, to cannibal ghost, to prey animal.

LIFE CLASSIFICATION

Meta-affinal siblingship is the basic mode of Guajá descriptions of their relationships to non-Guajá Amerindians, plants, animals, and supernatural beings. Most communities stand in the *harəpiana* maternal siblingship relationship to the Guajá. However, some communities are viewed as having a closer relationship to the Guajá and the consanguineal *harəpiháro* paternal siblingship is used to describe these relations. The paternal siblingship term is used to refer to the Guajá people as a whole, Guajá relationships with their spirit doubles, relationships with non-Guajá spiritual siblings, and the relationship of the Guajá community as a whole to the howler monkey community. Egalitarianism is key in understanding their perceptions of others, for it is the siblingship relation of social equivalence that is applied to other beings.

Being Awa *and the* Awa Being

The term *"awa"* roughly translates as "people" or "men" and is a group-referential term as well as being used in opposition to the term *kwɨya* (women). Within the category of *awa* there are distinctions made between the *awa-te* (true Guajá) and *awa-mihúə*[2] or *awa-parenči*[3] (different Guajá). *Awa-mihúə* and *awa-parenči* are terms used synonymously to describe those who either have been recently introduced to the Guajá village from other areas due to loss of territory or those who live deeper in the forest without contact with the FUNAI post. The otherness of the *awa-mihúə* introduced into the group is often marked by description of their distant origin and their dietary differences, and although they are incorporated into the group through marriage, their outsider status is often reinforced through teasing or scapegoating. For example, the source of a viral respiratory infection among the group was attributed to an *awa-mihúə*, and the birth defect a child was blamed on the eating of capybara by another *awa-mihúə*. Although capybara is taboo for the Guajá, the Guajá frequently break their food taboos, and it seems that the *awa-mihúə* are prejudicially targeted as responsible for mishaps occurring in the group.

The uncontacted Guajá living in the forest are not considered a unified group, but go by different designations depending on the group's relationship to a given individual on the reserve. In one recently contacted family, three of the adolescents married into the village group, and their forest group is often considered to be *awa-te*. Other groups are less well-known by the Guajá and are called *awa-mihúə*. In addition, there are some groups referred to as *aramaku* who have hostile relations with the village group, and most of the Guajá do not consider these individuals to be *awa* at all.

The Guajá self is perceived as both a singular and a plural concept (see also Cormier in press). Singularly, an individual consists of a body, *ipatérə*, and a spirit, *hatikwáyta*,[4] which represent a form of mind-body or spirit-body dualism in that the *ipatérə* is considered separable from the *hatikwáyta* in dreams, ritual, and death. Plurally, the Guajá believe in the simultaneous existence of multiple past versions of themselves, which are also *hatikwáyta*, which lead independent existences in the *iwa*, their celestial home.

While the *hatikwáyta* is considered to suffuse the physical body, its greater importance is in its representation of past images of the self. It can perhaps be described best as the "remembered self," and it is the image represented in dreams, memories, and photographs, but not in reflected images. Mirror images reflect the individual in the present time, but the images of dreams and memories are interpreted as existing outside of mundane world, outside the here and now and, thus, are considered to be spiritual. Photographs then, like memories, are recognized as representations of the past and are considered to be *hatikwáyta* images.

The Guajá *hatikwáyta* conform to the classic description of animism by Tylor (1958 [1871]) in its correlation with the duality of the self in the dream experience, whereby the self is believed to exist simultaneously in two different places. Durkheim's (1995 [1912]) critique was, in part, that it was difficult to reconcile this with the content of dreams, which often contained material from the past. It was illogical for him that any one would believe that they were both in the past and the present. For the Guajá, dreaming does involve a simultaneity between past and present, which is discussed further below.

In addition to the *hatikwáyta*, there is another spiritual aspect to the Guajá individual called the *aiyã*, which comes into being at the moment of bodily death. Upon death, the *hatikwáyta* separates from the body and goes to the *iwa* to remain forever, while a malevolent, earthbound spirit is created which eats the living. While all forms of life considered kin have *hatikwáyta*, *aiyã* is characteristic only of the *awa*. The *aiyã* are cannibals that prey upon the Guajá, eating both their *ipatéra* and *hatikwáyta*, and they are considered the ultimate cause of most pain, illness, and death. The *aiyã* are said to stay with the dead body during the day and then roam near the village at night, where they are frequently seen by the Guajá and are the source of much worry and concern. The *aiyã* are said to look exactly as the person did in life. If an individual's death has been recent, the *aiyã* is referred to by the same name as the individual had in life. The recently created *aiyã* are the ones usually blamed for a specific illness in an individual. After a few years, there is a genealogical amnesia for specific *aiyã* names, and they become part of a generic category of the *aiyã* in the same way that the ancestors become a generic category in the *iwa*.

The Spirit Siblings

Each Guajá is named for a plant, animal, or object with which the individual has a special spiritual connection. Specifically, they are called *haĩma* but are also considered to be *harəpihárə*. For example, the individual Takamã-či'a was named for his *haĩma*, the tukumã palm, *takamã*. When children are born, they are not given names, but rather it is up to the parent, and usually the social father (that is, the mother's husband), to discern the name, discovering which form of life or environmental feature is the child's *haĩma*. The name is usually determined by the time the child is several months old.

In some respects, the *haĩma* relationship resembles individual totemism. Individuals are considered to have a spiritual connection with a community of others. However, there are no taboos on the eating a member of one's *haĩma* community, and generally, they may be foods that the Guajá eat, foods that they do not eat, or objects that are not food at all. Of the *haĩma* determined, 64 percent were animal spirits, 19 percent were plant spirits, 6 percent were spirits of environmental features, 6 percent were spirits of manufactured goods, 2 percent were honey spirits, and 2 percent were spirits of the divinities. Although the *haĩma* typically represented useful species, they did not necessarily represent the most useful of the species. For the animal species, 43 percent were bird spirits, 31 percent fish spirits, 10 percent reptile spirits, 8 percent mammal spirits, and 8 percent insect spirits.

I suspect this phenomenon may bear a historical relationship to Amazonian shamanism. However, among the Guajá, there is no specialized shamanic role, and all individuals have *haĩma* siblingship. In addition to human beings, spirit *harəpiháro* are also attributed to some divinities and monkey pets. The divinities often have cannibalistic relations with their spirit *harəpiháro*. This will be discussed further, but at this point, suffice it to say that cannibalism and totem taboos are antithetical relations. Also, that spirit *harəpiháro* were identified for some of the monkey pets further supports the idea that nonhuman forms of life are seen as having a social equivalence to human forms of life, which suggests that the *haĩma* is best interpreted as animism rather than totemism.

The Forest and Sky Siblings

The *harəpiana* relationship reflects the equivalence of plant and animal communities to the Guajá community as a whole. The *awa* community, both living and dead, are referred to collectively as *awa harəpiháro*, who are opposed to the other communities who stand in the *harəpiana* relationship to the Guajá community. Two main classes of *harəpiana* communities exist, the *iwarəpiháro* and the *ka'arəpiháro*, which involve both spatial and temporal distinctions.

The *iwarəpiháro* are the inhabitants of the sacred sky, the *iwa*, and the category comprises the divinities, the *karawa*, who are in the *harəpiana* relationship, and dead ancestors and past versions of living individuals who are called *awa harəpiháro* and *awa harəpiháro-tikwáyta*. Although it may seem contradictory that the Guajá sky ancestor community is referred to through consanguineal siblingship but are more generally affinal siblings at the higher, more inclusive *iwarəpiháro* node, the scheme conforms to Viveiros de Castro's (2001) GUT theory, where consanguinity is hierarchically nested within affinity.

The dead, past selves and past others share the feature of being past mental images, but for the Guajá, these remembered others do not reside in the mind, but in the physical place of the *iwa*. The other inhabitants of the *iwa*, the *karawa* divinities, are believed to have always existed in the *iwa*, with the exception of the creator God, Mai'ira, who is believed to have lived on earth among the Guajá at one time. The *karawa* are benevolent, and the Guajá petition them for various favors, including healing and providing game.

The *ka'arəpiháro* (or *irarəpiháro*) are the forest kin, and the category includes plant life, animal life, and environmental features such as rivers and rocks. As Balée (1994a) described, among the Ka'apor and Guajá, plant life is divided into three major categories: *wɨra* (trees), *ka'a* (herbs), and *wɨpo* (vines). The term *ka'a* also refers to the forest in general. Introduced domesticated plants such as manioc and corn, while considered to be *ka'a*, are not considered *ka'arəpiháro*.

That is, they are not kin to the Guajá in the way that the nondomesticates are. Interestingly, the Ka'apor refer to the Guajá as *ka'apehar*, which Balée (1984:95) translates as "people of the forest"; the term appears to be a cognate with the Guajá term for forest kin.

Some forms of animal life are broken down into additional categories, but others are simply subsumed under the general *ka'arəpihárə* term. Fish are classified under the cover term *pira*. Birds are distinguished by size into the categories of *iramiri*, typically small, arboreal, passerine songbirds, and *ipopóə*, the larger arboreal birds such as curassows and parrots. Parrots have a separate cover term, *arárə*. Snakes are covered by the term *inama'ĩ*; however, some Guajá do not classify anacondas and boa constrictors as snakes because they do not bite. Additionally, some Guajá consider the Guajá living in the forest, as well as the *aiyã*, to be *ka'arəpihárə*.

Howell (1996) described a similar lack of higher-order differentiation in the classification of the Chewong of the Malaysian tropical forest. He described them as having a scheme predicated on identifying and naming rather than on clustering, and the underlying ordering principle was equality rather than hierarchy, which he related to the their sociopolitical order (Howell 1996:131). Guajá individualism and lack of social hierarchy (excluding some age and gender differences) may also relate to their classification scheme, but the essential social equivalence applied to nonhuman forms of life likely also plays a role in the lack of differentiation.

Although the Guajá do not have a category for monkeys, many now will use the Portuguese term *macaco* to refer to the class. The kinkajou (*Potus flavus*) is included by some as a *macaco* because of its prehensile tail. However, those that regard the kinkajou, *yapuči*, as a monkey agree that it differs from the other monkeys in that they are not able to keep it as a pet. The Barí of Venezuela also lack a generic term for monkeys but similarly recognize the Spanish term *mono* for monkey and also include the kinkajou and the closely related olingo (*Bassaricyon gabbii*) (Lizarralde 1997, 2002).

Hanima

The *hanima* are the Guajá pets. The Guajá will keep almost any animal as a pet if it will live. During this research period, pets kept included monkeys, agoutis, pacas, peccaries, coatimundis, sloths, one jaguar kitten, a variety of birds, opossum, mice, armadillos, caimans, lizards, tortoises, grasshoppers, and ants. Typically, pets are acquired as young orphans after their mothers have been killed for food. They are then kept and nurtured by the Guajá and are never eaten. Although it is always infants and juveniles that are kept as pets, there is no taboo on eating young animals, and the Guajá often kill young animals as well as adults. Dogs, cats, and chickens are also kept, but these are classified somewhat differently, as will be discussed below.

The *hanima* can be considered a transitional category between the *awa* and *ka'arəpihárə*, standing between the forest and the people and between edible and inedible. The term itself is sometimes ambiguously used. When hunters bring back live animals, they will say that they will not eat them because they are *hanima*. However, when hunters retell a story of a hunt, they will sometimes refer to the hunted animal as *hanima*. Monkeys stand in the closest relationship to the Guajá in that they are incorporated into the kinship terminological system used for people, with terms of address and reference, whereas most other animals are not (see chapter six).

Kinship Relationships Among Plants and Animals

The same-sex sibling terms of *harəpihárə* and *harəpiana* are used to describe not only the relationship of the Guajá to other forms of being, but also to describe the relationships among forms of life in the natural and supernatural world. The Guajá either emphasize the maternal same-sex sibling term, *harəpiana-te*, or use the paternal same-sex sibling term, *harəpihárə*, to describe plant and animal communities that are considered more closely related to each other than just the general *harəpiana* relationship among forms of life.

Of particular significance is that the *awa* consider themselves to be *harəpihárə* or *harəpiana-te* with only one other form of life, the howlers. According to the Guajá, the similarity between them is that they both "sing," referring to the loud, extended territorial vocalizations of the howler monkey. In addition, as discussed in chapter six, the howlers are believed to have been created from the Guajá people. The egocentrism deriving from plural paternity is also expressed through siblingships among other forms of life. While forms of life are considered to explicitly share a father, partial consanguinity applies to the relationship among communities of other beings. For example, the Guajá are considered *harəpihárə* with the howlers and the howlers are *harəpihárə* with bearded sakis, but the Guajá and the bearded sakis are not *harəpihárə* to each other (see table 5.1).

TABLE 5.1 Siblingships Among the Forest Species

Guajá Name	Common Name or Scientific Name	Harəpihárə or Harəpiana-te
awa	Guajá	howler monkey
wari	howler monkey	Guajá, bearded saki
kwičʰúə	bearded saki	howler monkey
ka'i-hu	Ka'apor capuchin	brown capuchin
ka'i	brown capuchin	Ka'apor capuchin, squirrel monkey
yapayu	squirrel monkeys	brown capuchin, owl monkey
aparikə	owl monkey	squirrel monkey, tamarin
etamari	tamarin	owl monkey
akuč i	agouti	paca
kararahu	paca	agouti
arapʰate	red brocket deer	gray brocket deer
arapʰa'ə	gray brocket deer	red brocket deer
tiaho	white-lipped peccary	collared peccary
mətə	collared peccary	white-lipped peccary
yawárə	jaguar	tapir
tapi'írə	tapir	jaguar, capybara
kapiawárə	capybara	tapir
yapuči	kinkajou	opossum
mukúrə	opossum	kinkajou
tamakai	squirrel	mouse
awynia	mouse	squirrel, rabbit
tapečĩ	rabbit	mouse
a'ɨ-hu	two-toed sloth	three-toed sloth
a'ɨ-te	three-toed sloth	two-toed sloth
taminawa	giant anteater	collared anteater
taminawa'ĩ	collared anteater	giant anteater
kamičia-te	yellow-footed tortoise	red-footed tortoise
kamičia-tu	red-footed tortoise	yellow-footed tortoise
yakare	alligator species	yakare-čũ alligator species
yakare-čũ	alligator species	yakarẹ alligator species
mitõ	curassow species	razor-billed curassow
mitõ-pini	razor-billed curassow	mitõ curassow
wira-hu	harpy eagle	wiraho (various hawks and eagles)
wiraho	various hawks and eagles	harpy eagle
tatu-te	armadillo species	tatu-pe armadillo species
tatu-pe	armadillo species	tatu-te armadillo species
yaku-pe	P. superciliaris guan	P. pileata guan

TABLE 5.1 Siblingships Among the Forest Species (*continued*)

Guajá Name	Common Name or Scientific Name	*Harəpihárə* or *Harəpiana-te*
yaku-te	*P. pileata* guan	*P. superciliaris* guan
takánə	toucan species	*takayu, takánə, takánə-hɨ* toucans
takayu	toucan species	*takánə, takánə-hu, takánə-hɨ* toucans
takánə-hu	toucan species	*takánə, takayu, takánə-hu* toucans
takánə-hɨ	toucan species	*takánə, takayu, takánə-hɨ* toucans
uru	vulture species	*apita* vulture
apita	vulture species	*uru* vulture
inaya	inajá palm (*Attalea maripa*)	babassu palm
wã'ɨ	babassu palm (*Attalea speciosa*)	inajá palm
takamã	*Astrocaryum vulgare* palm	*yúə* palm
yúə	*Astrocaryum gynacanthum* palm	*takamã* palm
pinõwa	bacaba palm (*Oenocarpus distichus*)	açai palm
yahárə	açai palm (*Euterpe oleracea*)	bacaba palm
muki'a	piquiá tree (*Caryocar villosum*)	bacari tree
mukuri	bacari tree (*Platonia insignis*)	piquiá tree
ako'o	forest cacao (*Theobroma speciosum*)	cupuaçu
kipu	cupiaçu (*Theobroma grandiflorum*)	*Theobroma speciosum*

Non-Kin

The *karaí* are the non-Indians, and this term is used for nearby Brazilian fisher-farmers with whom the Guajá have strained and sometimes hostile relations. They are neither *ka'arəpihárə* nor *harəpiana*. As previously mentioned, my husband and I were referred to as *americanos*, the term the FUNAI used for us, which I believed the Guajá used to avoid the problem of classifying us in the same category as the *karaí*.

Domesticated animals, including dogs, cats, chickens, and pigs, are not considered part of the Guajá kinship world. Although the Guajá keep dogs, cats, and chickens as pets, which are commonly called *hanima*, they are properly *karaí hanima* and not considered to be *harəpiana*. These animals exist only in the present and do not have the *hatikwáyta* necessary to go to the *iwa*. Thus, they do not have remembered images of their past selves in the *iwa*, and they do not go to the *iwa* when they die.

The FUNAI introduced chickens to the Guajá with the hope of involving them in raising the animals as a food source. While the Guajá will eat chickens, the problem is that when these animals live in close contact with the Guajá, they become part of the group and become *hanima*, even if they are still *karaí hanima*. As *hanima*, and they are not killed for food. Kracke (1978) reported a similar reluctance among the Kagwihív (Parintintin) to eat introduced domesticated chickens and pigs because of their association with their pets, *renymbá*, which are not eaten.

The Guajá do not have a specific category that distinguishes domesticated plants from nondomesticated plants. However, they are clear that introduced domesticates such as manioc and corn, while *ka'a*, are neither *ka'arəpihárə* nor *harəpiana*. Like domesticated animals, the domesticated plants are excluded from the *iwa*. The Guajá do not eat domesticates in the *iwa*, and domesticates do not go to the *iwa* when they die as other plants do. It should be noted that while most informants were firm on there being no domesticates in the *iwa*, some of the teenagers told me that the *iwa* did have manioc, which likely reflects the changing lifestyle of the Guajá as domesticates are beginning to be incorporated into their cosmology.

Non-Guajá Amerindians are classified as *kamara* and specifically include the neighboring Guajajara, Ka'apor, and the uncontacted *"aramaku"* foragers living on the reserve with whom the Guajá have hostile relations. Informants varied in their classification of the *kamara*. For some, they were classified as *ka'arəpihárə* type *harəpiana*. Some called them *ka'arəpihárə* but said they were not *harəpiana* to them. For others, they were neither *ka'arəpihárə* nor *harəpiana*. My sense is that the varied responses are similar to the negation of social relations within the group among individuals who have strained personal relationships. It seems to be a kind of social distancing for some so that the *kamara* are more like the *karaí* in not really being a part of their kinship universe. Those who considered the *kamara* to be *harəpiana* were clear that they did not go to their *iwa* when they died but suggested that they might have another *iwa* different from their own.

The Guajá distinguish themselves from the *aramaku* in several ways. The *aramaku* tie their foreskins with tukumã palm fiber rope at all times, while the *awa* reserve this for their *karawa* ritual. The *aramaku* wear armbands and head-bands made from parrots and *hairə* (a kingfisher-type bird) feathers rather than the toucan and vulture feathers worn by the *awa*. The Guajá report that the *aramaku* make fire while the Guajá do not know how. Further, the Guajá say that they are unable to understand the speech of the *aramaku*. The reported linguistic differences could suggest that the uncontacted Indians are not closely related to the Guajá; however, the significance of this is unclear. From descriptions by the Guajá, most of the two groups' interactions involve shooting at each other and running from each other, so it is unlikely that much direct communication is taking place.

CLASSIFICATION OF SPACE AND TIME

Past, present, and future are not entirely linear concepts for the Guajá, and temporal states are bound up with both spatial loci and spiritual states (see also Cormier in press). For Western culture, experiencing the past is a mental activity occurring in this space and time with a locus in the "mind." But for the Guajá, in dreams and in the *karawárə* ritual they leave the physical earth and go to another place, the *iwa*, where they see the dead as well as past versions of themselves and others. In these contexts, remembering is a ritual and a physical activity rather than a purely mental one. This is certainly not to suggest that the Guajá are unable to remember past events outside of dreams and ritual experiences, but rather that remembering in these contexts is conceptualized as having a physical reality.

The Classification of Space

The Guajá recognize four basic vertical divisions of space: *wɨ* (earth), *ka'a* (forest), *iwa-čũ* (light-colored sky), and *iwa-pinahũ* (dark-colored sky). The *iwa-čũ* and *iwa-pinahũ* are not divisions of the daytime and nighttime sky, but rather are considered to be separate sky layers, with the *iwa-čũ* the closer sky that is inhabited by the birds, the sun (*kwarəhɨ*), the moon (*yahɨ*) and the stars (*wə-wə*) and gives way to the *iwa-pinahũ*, which is very distant and is inhabited by the *karawa* divinities, the dead, and past versions of the Guajá. The distinction is likely related to the fact that the two most important spiritual activities of the Guajá, dreaming and the *karawárə* ceremony, both occur at night. However, in common speech, no distinction is made between the *iwa-čũ* and *iwa-pinahũ*, and both are referred to as *iwa*.

Some Guajá also describe a parallel world below them. Beneath the *wɨ* (earth) exists another *iwa-čũ* with another *ka'a* and *wɨ* which is inhabited by the same types of life-forms that live in the world above, including other Guajá, their plants and animals, and other *kamara*. Some use the term *manio'i* to describe the *karawa* divinities living in the parallel world below. As previously mentioned, the *iwa-pinahũ* for the beings of the world below is the Guajá's *wɨ*.

The Guajá use relative terms and number of times slept overnight to describe distances. The terms *kwatɨte*, *pebe*, and *mimbe* are used to refer to relative distance and indicate respectively near, far, and very far. The term *kwatɨte* can mean anything from right next to you to a ten minute walk. Distances that are very far are described in terms of how many nights the Guajá *kere* (sleep), which indicates that they have walked all day and then spent the night away from the village. The Guajá only have three numerical terms, *makamatũ* (one), *inahũ* (two), and *haí* (many). However, they will sometimes use kinship terms to indicate number above two. For example, they may say *či* (brother), *čikari* (sister), *čipa* (father), and *amã* (mother) to indicate sleeping overnight four times. But these terms are not used in any particular order, and so each term likely indicates "another different one," rather than indicating a specific quantity.

The Classification of Time

One area where the Guajá's non-linear view of time is apparent is in their distinction between relative ages, with the terms *púə*, indicating younger and beautiful, and *imánə*, meaning older. The terms are used, for example, to describe an older sibling as *imánə*, and a younger sibling as *púə*. However, the Guajá describe the *iwa* as *púə*, while the earth and forest here are *imánə*. At first, this seems contradictory because the *iwa* holds the past forms. But, the *iwa* does not so much represent the past in a time line where life that has existed longer is older, but rather, the *iwa* contains the earlier versions, the younger and more beautiful forms of life, which would be *púə* relative to life and time on earth.

In addition, the Guajá will use relative age in order to locate events in time. For example, one informant, when speaking of events that occurred shortly after the creation of the Guajá world, indicated how long ago it was by stating that it was when a woman of about thirty-five years of age looked like her young daughter, who is about four years old. The comparison is also telling in that it demonstrates how the depth of their view of the past of their people is truncated.

The *iwa* is held to contain personal pasts rather than past events and is a place of sacredness rather than of history. The past forms are not historical forms because individuals are transformed into a sacred state when they move from the world of the here and now on earth to the world in the sky. Past selves are enhanced in the *iwa*, where everything is good and beautiful, and there is no pain or suffering. For example, there was a three-year old child who survived a snake bite the previous year. The father stated that although younger versions of his daughter exist in the *iwa*, none of them had ever been bitten by a snake. Historical events are conceptualized differently than the permanent past forms of life that live independently in the *iwa*.

It is tempting to compare the simultaneous, independent existence of multiple versions of the self in Guajá cosmology with Everett's many-worlds interpretation of quantum mechanics which posits infinite parallel universes (see Gribbin 1984). However, the Guajá conceive of only one *iwa* that houses the past life-form versions. In addition, although logically a separate past life-form could exist for every past point in time, it would be a mistake to interpret their experience in this linear way. While they do say, "*Yaha haí iwabe*" (there are many of me in the iwa), these images of past selves and others do not have an a priori existence; they are generated from the experiences of memories and dreams rather than from a principle of infinite selves.[5]

The principle of transformation is central to the *iwa* life-forms. In the *iwa*, everyone is said to look like the divinities, with beautiful headbands and armbands and feathers on their bodies, as is the appropriate dress for the *karawárə* ritual. Everything is enhanced in the *iwa*. Men and women alike are beautiful, healthy, and very fertile. One informant described the children of his own *hatikwáyta*, which included forms of his children here on earth as well as additional children he fathered in the *iwa*, although he did not know the names of the *iwa* children without counterparts on earth.

The physical place of the *iwa* is an enhanced version of the of the earth. Many of the Guajá describe the forest on earth as being ugly, *manaheĩ*, while the *iwa* is described as being beautiful and flat, with some informants describing only a small forest in the *iwa* and others describing a large forest, but its being far away. The small forest was described as preferable because it is easier to hunt the monkeys in trees. It is possible that the smallness of the forest, as well as being an easier form to manage, may suggest an immature forest, that is, an earlier form. In addition, some describe the trees themselves as transformed, with spiny palms becoming spineless.

The animals in the *iwa* are similar to animals on earth with a few exceptions. One difference is that the animals in the *iwa* do not bleed when killed; they are said to stay beautiful. In addition, several the Guajá describe an additional type of howler called *wariyu*, which is red. Several Guajá saw a picture of the red howler monkey (*Alouatta seniculus*) in a book of mine and described it as being the *wariyu* of the *iwa*. One informant stated that she had seen this monkey in the forest on earth. Although red howlers are not known to this area, a study by Schneider et al. (1991) described pelage variation of the red-handed howler (*Alouatta belzebul*) in the eastern Tocantins area as varying in color from reddish to completely black. So it is very possible that this represents a actual phenotypic variant of the red-handed howler. The howlers kept as pets by the Guajá demonstrated some pelage variation as well, with most red-handed to some extent, but others solid black.

Some animals are not present in the *iwa*. In addition to the domesticated animals, the *mukúrə*, which includes possums and skunks, do not go to the *iwa*. The *mukúrə* are associated metaphorically with the *aiyá* due to their nocturnal habits and the foul odor of the skunk. Snakes and insects are absent from the *iwa*, with the exception of anacondas and boa constrictors, which are not classified as snakes. A few animals that are not eaten do go to the *iwa*, including vultures, which are never eaten, and the capybara, which is taboo, although sometimes eaten.

The *iwa* is where the Guajá look in order to understand both their past and their future. Visiting the *iwa* during the *karawárə* ritual provides a glimpse into what the Guajá can expect in the afterlife. It should be considered that it may be a distortion to associate the *iwa* with the future at all. The *iwa* may reflect a "beforelife" more than an afterlife since the dead are transformed into earlier, more beautiful versions of themselves. However, the dead as well as the past versions of those still living are believed to have a continuing existence in the *iwa*, which does suggest that an idea of the future is associated with the *iwa*, although its meaning cannot be easily glossed with the Western conception of future.

Space-Time and Dreams

The *hatikwáyta* leaves the body under three different conditions: death, dreaming, and the *karawárə* ritual. The connection of the *iwa* with past life-forms relates to the connection of the *iwa* with memory and the dead, which is in turn connected with the dream state. Death represents the last of the *hatikwáyta* past forms as the last version of the individual separates from the physical body and continues its existence in the *iwa* in an enhanced state.

The Guajá now bury the dead, but this may be due to the influence of FUNAI. I was unable to find anyone who was able to remember how the dead were treated before the introduction of metal tools to the group, although some older individuals did remember receiving their first steel tools from FUNAI. As previously mentioned, the FUNAI records of their earliest contact with the Guajá (which unfortunately coincided with much death from introduced diseases) reported many corpses being laid along the bank of the river (V. Parise 1973b). This practice seems consistent with the description one informant provided of what eventually happens to the dead body. He reported that during the rainy season, the divinities send much rain which cleanses the dead body. Then they begin adorning the body with the traditional headband, armbands, and feathers, transforming it into a beautiful state. Then the body is blown upon by the divinities and flies up to the *iwa*. The early FUNAI reports of the Guajá's placing bodies near the water may relate to the role of the divinities in cleansing the body with water. In addition, if a body was placed near the river, it is unlikely that any trace of it would remain by the following dry season. What scavengers did not consume would be washed away with the seasonal flooding.

Dreaming, like death, takes one to the beforelife.[6] While it is an activity recognized as occurring in both men and women, it is interpreted differently by gender. When men sleep, their *hatikwáyta* travel to the *iwa* and interact with past forms. Dreaming presents some danger for a man, since he is having an experience similar to dying in the flight of his *hatikwáyta* from his body. An important responsibility of a wife is to hold her husband in the hammock and to sing during the night as a kind of homing device for the men so that they can find their way back to the earth. It is quite a remarkable custom, since periodically all through the night, various women can be heard singing in an extremely loud, high-pitched soprano voice, and it is amazing that anyone could manage to sleep at all. When in the *iwa* during this dream state, the Guajá men interact with the divinities, past versions of themselves and others, and with the dead Guajá. While the dreams of women are also spiritual experiences, they are interpreted differently. The most important difference is that women never actually go to the *iwa* until they die. They are either able to see the *iwa* through the eyes of one of their *hatikwáyta* in the *iwa*, or they are spirit-possessed by a divinity and able to see the *iwa* through the eyes of the divinity.

The Guajá do not seem to recognize nightmares in the same way that members of Western societies do.[7] While they do describe *pinahi*, when a child wakes in the night in fear, this is an experience considered to be limited to children. The Guajá stated that they did not know what specifically caused children to be afraid at night but attributed it to immaturity, and they interpreted this *pinahi* as something completely different from *iwa* dreaming. Neither young girls nor young boys go to the *iwa*, and their dream experiences are interpreted in the same way as that of women.

The dream experience appears to be the basal experience on which the spiritual life of the Guajá is founded. The *karawárə* ritual is, in a sense, a waking dream or a dramatization of the dream experience, where it is reenacted in a conscious and controlled way to meet specific ends. The behavior displayed by participants resembles sleepiness more than a trance state. The ritual is in a sense all-purpose, in that it is the ceremony where healing takes place, people from the past forms of others and dead ancestors are visited, requests are made for success in hunting, and marriages are decided.

The *karawárə* ceremony takes place in the evenings and involves the entire community. The men dress in headbands and armbands made from toucan feathers, apply vulture down to their bodies, and wear the traditional *yaměi* tukumã palm–fiber string around their foreskins. As stated before, this specific dress makes them appear like the *karawa* divinities as well as the dead and past selves who dress in this manner in the *iwa*. Past forms of women also dress in feathers in the *iwa*, although the women on earth do not wear feathers for the ceremony.

The *tačai* hut is another important element of the ceremony. It is a small thatched structure just big enough for one person to stand inside. The men take turns entering the *tačai* during the ceremony, and when they exit they burst through the door as though they had just landed back on earth. While the *tačai* provides a transitional space or a portal between the earthly world and the *iwa* and solves the visual contradiction of seeing the individual when he is supposed to be in another place, the *tačai* is not essential for going to the *iwa*. Some men leave and go to the *iwa* without entering the *tačai* at all. So, while it is a tool that probably is very useful in achieving the *iwa* experience, it is not essential to it.

Although only the men actually travel to the *iwa*, the women and children are also involved in the ceremony. The wives of the men assist in applying the feathers for the ceremony and sing throughout the ceremony, as do the men. Although this is a sacred ceremony, the atmosphere is surprisingly casual. Although children also sing, they also run around and between the dancers playing and laughing, with young boys often imitating the men.

Spirit possession also takes place during the *karawárə* ritual as the divinities enter the bodies of the men while their spirits are in the *íwa*. While possessed, the men will blow air on others to heal them or aid them in some way. For example, a man may blow on his wife to relieve pain or fever or to assist her in finding tortoises, which are associated with female fertility. Only the *karawa* are involved in spirit possession. Dead ancestors are said to be unable to ever return to earth because it causes them great physical pain.

During the ceremony, the Guajá petition the divinities for game animals. Two types of divinities are petitioned, the masters, or controllers of animals, and the mothers of animals. For example, in the case of howlers, the Guajá ask Yu-či'a, the master of the howlers, or Wari-ihɨ, the mother of howlers, to bring the monkeys closer to them. A woman can receive assistance in collecting tortoises indirectly through her husband, who, when possessed by the female tortoise divinity Kiripi, blows on his wife during the *karawárə* ritual.

Healing is received in several ways from the divinities. An individual male can directly obtain healing for himself from a divinity in the *iwa*, he can heal others through spirit possession, or there can be a general petition for healing the group through the wind. The *karawárə* rituals begin at the beginning of the dry season when a blustery wind begins to blow. This is interpreted as the work of the divinities to heal the Guajá. The dry season may be a healthier time in general for the Guajá as the disease-carrying mosquitoes diminish and the humid conditions that exacerbate introduced viral infections diminish.

Another important function of the *karawárə* ritual is deciding first marriages. If a man wants to marry, he goes to the girl's father and asks him. The father will then go to the *iwa* and consult the divinity Araku. This means that the marriage of young girl tends to be arranged during the dry season when the *karawárə* ritual is performed. In addition, consulting Araku is a social leveling mechanism; the authority for saying yes or no to a marriage is transferred from the father to the divinities.

Remembering the dead in the *iwa* experience explains both death and dreams. It serves as a way to explain what happens when the Guajá die and also helps diminish the pain of losing a loved one. Men are able to make conscious decisions to visit specific individuals during the rituals, and women can get information about dead relatives second-hand from the men. Women may also have the opportunity to see dead relatives in their dreams. The cosmological beliefs explain and reinforce social status differences which are based on three dimensions: gender, age, and marital status. Although it is often hard to see status differences in gender in the day-to-day life of the Guajá, the exclusion of women from the *karawárə* ritual is rather dramatic, as it symbolically excludes women from access to the past and from access to the divinities. Lack of access to the divinities is a major limitation of power. As mentioned previously, the Guajá do not believe that women can actually go to the *iwa* in their dreams; they are either spirit-possessed or see through the eyes of their *hatikwáyta*. This makes female encounters with the divinities passive, while male encounters are active. It is interesting to compare this with the structuralist research of the domestic/public sphere dichotomy that was popular in gender and anthropology in the 1970s (e.g., Chodorow 1974, Ortner 1974, and Rosaldo 1974). One of the major criticisms leveled against the cultural association of women with a domestic and interior space and men with a public and exterior space is that such a construct was inappropriate for many nomadic peoples, such as hunter-gatherers, where a permanent women's space simply did not exist (e.g., Leacock 1978 and Lamphere 1993). The Guajá, however, do practice gender segregation in their spiritual space, if not in the forest itself, as women are confined to the earth while men have access to the earth and the sky, the mundane and the sacred.

To a lesser extent, the *karawára* ritual reinforces differences in age and in marital status, which are somewhat related concepts. Although there does not seem to be a hard and fast, prescribed rule, going to the *iwa* is reserved for the adult married men of the group. A man should be married when he goes to the *iwa* so that his wife can sing to help him find his way back to the earth. Graham (1995) somewhat similarly describes the dream experience as being interpreted differently among the Xavante depending one's age grade and status. Among the Guajá, when a young man first goes to the *iwa*, it is not seen as an inherent right he acquires through a rite of passage into adulthood; rather, it is understood as more of an achieved ability developed through maturation. So young boys are not so much prohibited from going to the *iwa* as they lack the ability to do so. Typically though, it is not the young man himself who makes the decision to first travel to the *iwa*, but someone else in the group who encourages him to go. The Guajá men described their fathers, fathers-in-law, and even women in the group as being the ones to encourage their first trip to the *iwa*. Although individualistic, some social leveling seems to be involved, as the decision for the first trip is encouraged by someone else in the group.

The *iwa* acts as a cultural filter (see also Cormier, in press). Although life in the *iwa* is enhanced, it does not include items that are not traditionally part of the Guajá's culture, even if these items are extremely useful or highly desired by the Guajá. As mentioned previously, the Guajá are growing increasingly dependent on the domesticated plants that have recently been introduced. However, domesticated plants and animals do not exist in the *iwa*. The Guajá also view many Western items that have been added to their lives as extremely useful, such as metal tools, Western clothing, and guns, but none of these appear in the *iwa*. Interestingly, the Guajá say that there are *awa* in the *iwa* who know how to make fire, although the living do not know how to make it themselves. It is possible that the *awa* of the *iwa* represent a historical memory of a time when the Guajá possessed the skill.

Among other Amazonian groups, there is some evidence for ritual and spatial segregation of time. Among the Tukanoans of the Northwest Amazon, a ritual is enacted wherein the longhouse is transformed into ancestral time and men are transformed into ancestral people (Hugh-Jones 1979, Jackson 1983). While there is similarity to the Guajá in assigning temporal states to physical locations, the Tukanoans access the past by transforming the same space, including that of their own bodies, into the past. For both the Tukanoans and the Guajá, the past occupies a definite, limited physical place that can be accessed through ritual.

The Wari' (Pakaa Nova) have similar animistic beliefs regarding dreaming and the soul. During dreams, the soul is believed to leave the body and wander (Von Graeve 1989). There is similarity as well in that individuals seen in dreams are believed to be the dead (Vilaça 1992). Further, Vilaça describes the idea of *jam*, which is an image of the body that has an independent existence and appears in shadows, dreams, and photographs. Similarities with the Guajá are apparent in the idea of a spiritual form of the self having an independent existence and the ability to see the dead in the dreams. Despite these similarities, the Wari' do not seem to associate temporal states with spiritual states in the same way as the Guajá do.

Seasonal Change and Perception of Health and Illness

Temporal seasonal changes and varying climatic conditions of wind, rain, and temperature are closely related to Guajá ideas about health and illness. Seasonal changes and weather events are in the realm of the divinities. Wind and rain are largely controlled by Tapánə,[8] the rain and thunder divinity, although all the *karawa* are said to contribute to rain and wind to some extent through their singing. Temperature is also closely linked to spiritual matters and, thus, to temporal states. Temperature, rain, and wind will each be discussed in turn in their relation to ideas of health, illness, and cosmology.

The Guajá's basic distinction between *haku* (hot) and *hačə'ə* (cold) is related to spiritual states: heat is associated with the dry season, healing, and the divinities, and cold is associated with the wet season, illness, and the *aiyã*. The sensation of cold and the experience of fright are both linked to the *aiyã*. The *aiyã* cannibal ghosts prey particularly on children and wait in hiding to grab them or blow on them causing them to fall to the ground shivering with fear. The high fevers that accompany malaria are considered to be *hačə'ə* (cold), rather than hot, due to the shaking and chills which are characteristic of cold and fright, both of which are associated with the *aiyã*.

The Guajá have incorporated Western knowledge into indigenous knowledge in their beliefs about malaria transmission. The FUNAI nurses have taught the Guajá that malaria is caused by mosquitoes, and now some Guajá describe mosquitoes as *aiyã*. Otherwise, the Guajá believe the *aiyã* to be a strictly human phenomenon with no other forms of life leaving an earthbound spirit when they die. It should be mentioned here that the Guajá conceptualize the frequent viral infections that they suffer from introduced pathogens differently than the *aiyã* diseases. Viral infections are referred to as *tata*, the word for fire, and are believed to be caused by smoking, which they have seen practiced by the neighboring Guajajara and other Brazilians. They say that the Guajajara and others eat fire, which makes them sick with *tata*, and the winds then spread the disease to Guajá. The fevers from these viral infections are typically called *haku* (hot). However, they will call them *hačə'ə* (cold) if the fevers are high enough to cause the shaking chills.[9] Although there is a distinction between Western and *aiyã* diseases, there is some overlap, or at least uncertainty in some cases, as to the ultimate cause of disease. However, any disease severe enough to cause death is attributed to *aiyã* cannibalism. Taylor (1996) has argued that in Amazonia generally, death is not viewed as resulting from natural causes, but from malignant human agency.

Related to the association of the *aiyã* with cold is their association with darkness. The *aiyã* only come out at night and are particularly afraid of fire due to its properties of heat and light. The Guajá try to keep a fire going all night because if it goes out they are then vulnerable to *aiyã* predation.[10] While the *aiyã* are associated with cold, the *karawa* are associated with heat. The *karawa* ritual is only performed during the dry season when the temperature during the day is warmer. The Guajá say they know the time is right for the ritual when the lizards start coming out to bask in the sun. Both the *karawa* and the *iwa* are thought to be hot.

Rain has a dual nature, seen as both good and bad. As mentioned before, the Guajá believe the rain is used by the *karawa* divinities in order to purify and beautify dead bodies before they are taken up into the *iwa*. The Guajá say that they do not perform the *karawárə* ritual during the rainy season for two reasons. First, the ritual involves spirit possession, when the *karawa* come to earth. During the rainy season, the *karawa* are believed to be already here, cleansing and purifying the dead bodies that have yet to be transported to the *iwa*. Another reason they give is that, literally, the feathers they apply for the ceremony will get wet, and the men will fall back to earth.

The wind, *witu*, is a transformational force with both the power to heal and to traverse temporal realms. Like the rain, it has a dual nature. When the *karawa* divinities blow, they can heal the sick, transform the dead, and transport the dead to the *iwa*. Singing is closely akin to the blowing of the wind, and singing by the *karawa* divinities has the same type of power as does their blowing. However, the breath of the *aiyã* is foul-smelling and harmful.

Inhaling odors is also closely related to the wind and healing powers. The vast majority of the medicinal plants used by the Guajá are strong smelling (*ka'akača'ə* or *irakača'ə*). Typically, bark or plant stems are soaked in cold water and then splashed on the body as a medicinal bath. Sometimes the Guajá will just wrap a piece of bark around their waist, head, arm, or leg if that area is afflicted. The foul-smelling *aiyā* are repelled by the medicinal odors and will run away in fear when they smell them. The Guajá use these treatments frequently on their children, but adults also use them. They also will apply the treatments to their pets.

RITUALIZED REMEMBERING AND GENEALOGICAL AMNESIA

The possibility exists that Guajá genealogical amnesia bears some relation to their conceptions of space-time, dreams, and memories (see also Cormier 1999, Cormier in press). Here, I am making somewhat of a Sapir-Whorfian argument in suggesting that the ritualization of remembering bears a relationship to Guajá genealogical amnesia. For the Guajá, remembering dead ancestors is not a mundane activity, but is a sacred activity occurring in dreams and in the *karawárə* ritual. While it certainly cannot be argued that the Guajá do not remember outside of sacred contexts, it is possible that the association of memory with ritual to some extent influences the way in which the Guajá structure memories or, at the very least, influences the way that they talk about memories and thus may relate to their extreme genealogical amnesia.

Although dead ancestors are important in the Guajá belief system, the nature of their ritual association itself may inhibit their recall. The temporal distinction between the sacred past and the mundane present situates dead ancestors outside of day-to-day kinship interactions, which is consistent with genealogical amnesia and the exclusion of ancestors from kinship reckoning. Interestingly, among the Mehinaku, the response to genealogical questions is often, "I'm not from mythic times" (Gregor 1985:62), which is similar to the Guajá response of "she went to the *iwa*." What is being communicated is more than a simple lack of knowledge. The Guajá are communicating that the question uses an inappropriate context, that is, "I do not know; that person is in the place of the remembered dead, while I am in the earthly present."

As past, present, and future are not entirely linear concepts for the Guajá, temporality is bound up with both spatial loci and spiritual states. In Western culture, experiencing the past is a mental activity occurring in the present space and time with a locus in the "mind." But for the Guajá, in dreams and in the *karawárə* ritual, they leave the place of the physical earth and go to another place, the *íwa*, and see the dead as well as past versions of themselves and others. In these contexts, remembering is a ritual and a physical activity rather than a purely mental one. Again, this is certainly not to suggest that the Guajá are unable to remember past events outside of dreams and ritual experiences, but rather that remembering in these contexts is conceptualized as being appropriate to these ritual contexts, which are viewed as an alternative physical reality.

Viewing the past as having a physical reality is not unique to the Guajá people. In lowland South America, among the Kagwahiv (Parintintin), the dream experience includes visiting dead relatives and is sometimes considered to represent reality (Kracke 1978). As previously mentioned, the Tukanoans of the Northwest Amazon reenact a ritual where the longhouse is transformed into ancestral time and the men are transformed into ancestral people (Hugh-Jones 1979; Jackson 1983). Here, the past occupies a definite physical place that can be accessed through ritual. In addition, the previously mentioned Warí (Pakaa Nova) concept of *jam* is an image of the body that has an independent existence that appears in shadows and in dreams (Vilaça 1992). There is a suggestion of the past as a physical reality here, since individuals seen in dreams by the Warí are also believed to be the dead.

Outside of lowland South America, some similarity exists to the dream-time of the Mardu Australian Aborigines. According to Tonkinson (1991), the dreaming, or dream-time, is of a past time when the movements and adventures of the Dream Beings shaped the physical landscape of the Mardu. Although the Dream Beings are now gone, having metamorphasized into celestial bodies or terrestrial features, their life forces remain and continue to affect the daily lives of the Mardu. The Mardu access the residual power of the Dream Beings through ritual and dreams. In dreams, the dream-spirit of a person takes the form of a bird, leaves the body, and travels to where it may encounter a spirit being. The spirit beings are intermediaries between the residual power of the Dream Beings and the Mardu people. In addition, spirit essences of the Dream Beings exist throughout the physical environment of the Mardu, including some in the form of spirit children that enter women to be reborn as human beings.

Parallels exist between the Guajá and the Mardu in their non-linear concepts of time, animism the inherent in their dream experiences, and bird symbolism. For both the Guajá and the Mardu, the past is not viewed strictly as specific events on a time line. Rather, past forms have a continuing existence accessed through the dream experience and through ritual. In Mardu dreams, access to the power of the Dream Beings is indirect and is attained through intermediary spirit beings that are able to bridge not only the human world and spirit world, but also past and present. Herein perhaps is the key difference between Mardu and Guajá conceptions of the past. For the Mardu, the past seems to perfuse the present in the continuing essence of the Dream Beings throughout their environment and in the transformation of spirit essences into human beings. The Guajá, however, have spatially segregated the past and present into physical places. Rather than the indirect transfer of spirit essences in Mardu dreaming, the Guajá go to the place of past beings and directly petition the *karawa* or interact with dead relatives.

Temporal and individual ambiguity are characteristic of both Guajá kinship and religion. Neither notions of the self nor of one's kinship relations are definite and clear-cut. Individual identity includes a degree of nebulousness when multiple versions of an individual exist in the *iwa*, who are all leading independent existences. Similarly, the notion of descent is nebulous with multiple fathers through whom kinship links can be manipulated to form or dissolve social relations based on the needs of the current situation. Despite Guajá kinship reckoning through paternal consanguines, their multiple "biological" fathers, multiple past selves, and the spiritual bifurcation of the dead would be inconsistent with the idea of a shared common substance among members of a descent group. Genealogical amnesia is more than a lack of knowledge; it is a functional part of Guajá religious and social life. Questions about ancestors are not historical questions for the Guajá, but are of a spiritual nature.

Guajá animistic beliefs are central to the way they relate to other beings in the natural and the supernatural world and to way they perceive their own spiritual natures in the past and present. In the following chapter, the social and symbolic role of pets in household will be discussed as well as the consequences of pet keeping to the monkeys.

6

PET MONKEYS

Nonhuman primates' value to human groups often involves more than being merely a source of calories. A number of groups across various cultures are interested in monkeys because of their behaviors, particularly their social behaviors. Keeping monkeys as pets for companionship is a common practice among lowland South American Indians, including the Aché (Hill and Hawkes 1983), the Barí (Lizarralde 1997, 2002), the Huaorani (Rival 1993), the Kagwahív (Kracke 1978), the Matsigenka (Shepard 1997, 2002), the Mekranoti (Werner 1984), the Upper Putmayo River Indians (Hernández-Camacho and Cooper 1976), the Wayãpi (Campbell 1989), and the Yanomamö (Smole 1976). Primate pet keeping is also practiced among non-Indians, particularly in villages and towns in Amazonia (Mittermeier and Coimbra-Filho 1977; Mittermeier and Coimbra-Filho 1977; Moynihan 1976b; also see Wolfheim 1983); several villagers I met living near the Guajá reserve had monkey pets. Monkeys are also kept in groups outside of the Neotropics, such as the Mende of Sierra Leone, who consider monkeys to be particularly good pets for young children due to their playfulness (Richards 1993:146).

In some groups, monkey behaviors are exploited for profit. In Thailand, pig-tailed macaques (*Macaca menestrina*) and long-tailed macaques (*Macaca fascicularis*) are used to harvest coconuts and other fruits. Monkeys used in this way contribute significantly to the income of their keepers, as their labor may involve picking over one thousand coconuts a day (Sponsel, Natadecha-

FIGURE 6.1 Child with infant howler (*Alouatta belzebul belzebul*)

Sponsel, and Ruttanadakul, in press; Sponsel, Ruttanadakul, and Natadecha-Sponsel 2002). In Bali, monkeys have an important economic role in promoting tourism. According to Wheatley (1999:54), in addition to religious reasons (see chapter seven), monkeys are fed by locals to attract tourists; for example, over 57 percent of the diet of long-tailed macaques in the Ubud monkey forest comes from human provisioning. Wheatley (1999:55, 129) reports also that 78 percent of vendors advertise the monkeys as a tourist attraction, and many sell peanuts for tourists to feed the monkeys. Small (1994) has also described thieves training monkeys to snatch wallets and other valuables from tourists in Bali.

PET KEEPING IN AMAZONIA

Numerous ethnographers have commented on pet-keeping practices in Amazonia. However, until recently, little systematic research has been done to understand how pet keeping relates to the larger cultural framework. Fausto (1999) has offered a model that draws on the work of Descola (1994), which incorporates affinity and consanguinity, predatory cannibalism associated with warfare, shamanistic control of supernatural beings, and the adoption of children. Descola (1994:339) linked these practices through a structural homology:

affine : consanguine :: enemy : captive child :: prey : animal familiar.

Fausto generalizes the model through his concept of familiarizing predation. He describes the relationship between the pet and pet owner as adoptive filiation and argues that it is the structural equivalent of the relationship between a father and an adopted child. He describes both as prototypical relations of symbolic control in Amazonia. Further, these can be structurally linked to the relationship between a shaman and a spirit familiar and that of the killer and the victim in cannibal warfare. What is key for Fausto is that all of these relationships involve social reproduction. "Others" are needed to produce the identities of subjects within the group.

Erikson's (2000) approach differs in taking a psychological perspective on pet keeping. He views hunting and taming as two complementary aspects of the assimilation of animals by humans into society. Erikson views the hunting of animals as creating a conceptual discomfort in terms of the value of reciprocity in Amazonian societies. Pet keeping is a means to resolve this cognitive dissonance. He makes the point that the favorite type of pet is often also the favorite type of game, and that generally the subset of animals kept as pets matches the subset of animals hunted as game.

Taylor (2001) describes links among Jivaroan ideas regarding marriage, seduction, hunting, and taming. In Jivaroan *anent* (magical thought songs), seduction of women is expressed in terms of a hunter's pursuit of prey. She connects Jivaroan ideas regarding the taming of women in marriage to the taming of pets. The brothers-in-law who give sisters in marriage, allowing one to have children, are equated with the game brothers-in-law whom one kills and takes pets from. The structural linking of women to game helps to explain why women are considered to be the natural caretakers of pets and have a quasi-maternal relationship to them.

The Guajá seem to share some aspects of these models, but they do not conform precisely to them. Descola (1998) has criticized the psychological perspective, arguing that this cognitive dissonance is more characteristic of Western views of animal killing than Amazonian ones. Descola's point is very well taken. While I do not believe that cognitive dissonance is the primary reason for Guajá pet keeping, it does come into play. The Guajá do express some uneasiness regarding the ambiguous status of infant animals when a decision needs to be made as to whether they will be eaten or become companions. In addition, the Guajá conform to Erikson's idea that a group's favorite game is also the group's favorite pet.

Fausto's and Descola's formulas, which link affinity and consanguinity, predatory warfare, shamanism, and pet keeping seems far-reaching. The general framework of linking sociality, cannibalism, pet keeping, and hunting are apt for the Guajá, but some aspects are problematic. My suspicion is that this formula may have applied to the Guajá earlier in their cultural history, prior to their shift to foraging, and that it has been reconstituted in terms of cultural features associated with foraging. The Guajá do not recall a history of institutionalized intra- or intertribal warfare, nor do they have the role of warrior nor the specialized role of shaman. However, Fausto's key point—that the relationship of humans to Others involves social reproduction—does apply for the Guajá. But I believe that there has been insufficient attention to the other side of the equation of social reproduction—the importance of pet keeping in producing female identities.

Taylor provides a fascinating analysis of the Jivaroan equation of hunting with seduction and the taming of women in marriage with the taming of pets. While the Guajá share some similarities with the Jivaro, they differ in many respects. Related to Guajá notions of plural paternity is the idea that women are sometimes viewed as sexually demanding of men. Guajá men do have some authority over Guajá women, but it is weakly expressed in their relatively egalitarian society. However, Taylor's description of Jivaroan pet keeping as a quasi-

maternal relation rings true for the Guajá, and I believe that the maternal relationship is key to understanding Guajá pet keeping.

Similar to the Jivaro and numerous other groups, meta-affinity is the foundation for the Guajá's relationship to nonhuman beings. However, one important
way that the Guajá differ from the Jivaro and many other Amazonian groups is
that the prototypical other is not the enemy brother-in-law, but the companionate,
matrilateral same-sexed sibling. Structurally, the relationship of a mother to her
child is not that different from the relationship of a female to her pet, in that both
are nonconsanguineal relationships. Taylor's (1998) term "affinal consanguines"
may be the best description of the relationship between mothers and their children, matrilateral siblings, and owners and pets. The basis of the relationship of
the Guajá to nonhuman others is the type of affinity created through relationships
traced through one's mother rather than that of the meta-brother-in-law. In brief,
Guajá pet keeping is better described as the reproduction of mothering.

GUAJÁ PET KEEPING

Among the Guajá, although monkeys are clearly important as prey, social interactions with pet monkeys are also part of their daily life. Young monkeys kept as pets
are incorporated into the Guajá kinship system as the children of Guajá women.
Monkey pets are treated as dependent children and, as such, serve to enhance the
culturally valued image of the fertile female. Monkeys, as well as most other pets,
are brought into the Guajá community as orphans when their mothers are killed
for food. In some instances, mothers may become separated from their young in
trying to escape hunters. The Guajá climb trees and pick up abandoned infant
monkeys after a hunt. Lizarralde (1997, 2002) reported similar behavior among
the Barí of Venezuela, who describe capuchins as abandoning their infants when
they flee from hunters. The Guajá never eat pet monkeys, since these serve, to
some extent, as surrogate children for Guajá women (see also Cormier 2002a).

When infant and young juvenile monkeys are brought into the Guajá community, they are given into the care of a female Guajá. The fate of the monkey
is in the hands of the woman, who may decide that the monkey is to be eaten
rather than kept as a pet. Once the decision is made to keep the animal as a pet
though, it will never be eaten. Pet monkeys cling to the hair on the heads of the
women, as they would cling to the bodies of their monkey mothers. Like the infant Guajá, infant monkeys stay in constant physical contact with the "mother"
and are cared for like an infant—played with, sung to, bathed, breast-fed,[1] and
generally nurtured. Monkeys will also eat premasticated foods directly from the
mouths of women.

All seven species of monkeys known to the Guajá are kept as pets, and each of the ten households included monkeys during the study period. Several households, in fact, contained more monkey than human inhabitants (e.g., one household with eight monkeys and six people and another with twelve monkeys and ten people). Although a variety of other animals are kept as pets by the Guajá, none are nurtured as children to the degree of the monkeys. While all pets are called *hanima*, pet monkeys are referred to as *hamïmïrə* (my child) by their human mother.

The greatest contrast is between the treatment given to monkeys and to dogs. Dogs are treated with extreme cruelty and most are emaciated, physically abused, and constantly being shooed away. Their primary usefulness to the Guajá is in tracking down paca and agouti, but they may be fed little or none of the game they assist in locating. While monkeys may be fed directly from a woman's mouth, dogs tend to be fed indirectly; they scavenge for scraps—often the scraps dropped by pet monkeys. The Guajá do not seem to recognize that their physical abuse creates the viciousness in the animals and say that dogs are *nəkatuí* (bad). They explain that while all forest life is considered kin, dogs and other domesticates lack souls.

Although all forest life is considered kin, pet monkeys are incorporated more directly into the kinship system at the household level. They are given personal names and kinship names of reference and address modeled on the kinship terms the Guajá use for one another. While some kinship terms are identical to the ones used for the Guajá, some are modified versions that can perhaps be best thought of as kinship sobriquets. For example, a black-bearded saki was called *čikari* (sister or daughter), a howler was called *čipaiyu* (yellowish-red father), an owl monkey was called *čiaʔї* (little mother's brother), and a tamarin was called *iménə-etamari* (tamarin husband). The personal names of the monkeys, like the personal names of the Guajá, represent a form of animism, in that they are named for a plant, animal, or object with which they are believed to share a special spiritual siblingship. Giving monkeys personal names suggests that the Guajá view the spiritual nature of monkeys as similar to that of Guajá humans.

Monkeys as Female Body Art

The monkey, in its role as dependent child, enhances the image of female fertility in several ways. First, surrogate monkey children act to mitigate the perceived asymmetry in male and female fertility. Men are fertile not only throughout their adult lives, but cultural beliefs are such that men claim the children of any woman with whom they have sexual contact either prior to or during a pregnancy. Men, then, are believed to have far more children than do women in a culture where children are highly valued. Here, a clear difference exists between the Guajá and many polygynous systems where some men have large numbers of children while others are excluded, because almost all of the Guajá men are believed to father children, typically many children. Women take great pride in their monkeys; the monkeys' role as surrogates can also be seen in the practice of giving pet monkeys to women to breast-feed after miscarriages. One young girl was still lactating nine months after a miscarriage because she had been nursing three pet monkeys. Women beyond their childbearing years often have the greatest number of monkey pets.

Monkeys can be thought of as a type of body art that projects an image of fertility and, thus, sexual attractiveness. Pregnant bodies and lactating breasts represent feminine ideals for the Guajá. They describe all women in the *iwa*, the sacred sky-home, as pregnant, lactating, and caring for many small children. Although men and boys do interact with monkeys, they are primarily in the care of women, as are human infants. Two men in the village had pet monkeys that primarily clung to them, but the remaining eighty-eight monkey pets were mostly in the company of women. The association of women with pets is also supernaturally expressed through the female divinity Piraya, who carries multiple monkeys on her head. The divinity Kiripi keeps tortoises as pets. However, the past and sacred forms of the Guajá in the *iwa* do not keep pets. They are said not to need them because they have many children.

Monkeys can serve convincingly in the role of child due to their physical and behavioral similarities to human beings, particularly as infants. The Guajá explain this quite literally when asked why other pets are not treated in the same way as monkeys, saying for example, that coatis fall off your head, and fish cannot live out of water. However, they also believe that monkeys' suitability involves a special, natural relationship between women and monkeys; women are said to "know" the monkeys in a way that the men do not, and the monkeys "want" the women while they "fear" the men. In this way, the social roles of female child rearing and male hunting are underlined (both men and women are gatherers).

Monkeys and Childhood Enculturation

Monkeys also play an important role in the socialization of children learning childcare and hunting skills. Young girls typically have had experience in taking care of a number of infant monkeys long before they bear their first child. While girls, and boys to a lesser extent, also learn childcare skills by caring for younger siblings, the monkey provides several advantages for the learning process. First, it is mechanically much easier for a child to care for a monkey baby than a human baby, not only because the monkey weighs less, but because the monkeys cling to the head of the girls instead of having to be carried. Related to this is the level of dependency among primates compared to other species. While a girl can leave her baby tortoise or agouti alone, the clinging of infant monkeys simulates the level of responsibility she will have in caring for a future infant. Further, girls can have direct experience in caring for monkeys at a much earlier age than they can with human infants. Girls as young as two may begin carrying a baby monkey on their heads, and girls as young as five may be the primary *amã* (mother) of an infant monkey, meaning that the monkey clings to her throughout the day, leaving only to be breast-fed by the girl's mother or another lactating female.

For young boys, the monkey is most important in learning hunting skills. Keeping monkeys as pets in the household familiarizes young boys with the calls of monkeys and aids them in recognizing and discriminating these sounds in the forest. One young boy learned to imitate a howler monkey call before he spoke his first word. While female toddlers may begin to carry monkeys on their heads, male toddlers get toy bows and arrows as soon as they are able to walk. A boy's father will make him bows and arrows that gradually increase in size and quality as he grows older, until he is proficient in the adult weapons. On several occasions, young boys were observed taking practice shots at older pet monkeys that were past the clinging stage. Parents never reprimanded children for this behavior and on one occasion actively encouraged a young boy by singing a song about hunting howler monkeys as he practiced. That such efforts by young boys are encouraged, even at the expense of pets, reflects both the changing role of the older monkey (discussed below) as well as the extreme importance of learning to hunt in Guajá culture. Finally, young girls and boys both learn to recognize monkey calls, and although women generally do not hunt with the bow and arrow, their ability to recognize animal sounds on trekking expeditions also would be an asset to their male companions. Young boys also learn childcare skills through taking care of and playing with baby monkeys, although generally they have less involvement with monkeys than do the girls.

The Costs of Keeping Monkeys

Like Burmese neck rings or Euro-American high-heeled shoes, monkey body art is practiced despite the cost of physical impediment. The Guajá incur considerable cost by keeping monkeys in terms of the food burden on the group, the sheer weight of carrying them on the trek, and the maturational changes that result in monkey destructiveness and physical aggression. Physical aggression, as will be discussed in the next section, is also detrimental for the monkeys who are kept in captivity by the Guajá.

The monkeys, as well as the other numerous pets the Guajá keep, create a burden on the group's food supply, directly through the need to procure a greater quantity of food and indirectly through the need of women who nurse monkeys to increase their caloric intake. While one monkey may not create a noticeable burden, typical households have multiple monkeys that are being nursed or provided with food and water. As previously mentioned, some households have more monkey pets than human inhabitants and often have a variety of other pets, most commonly agoutis, tortoises, coatis, and birds. In addition, the Guajá must also deal with waste disposal for their multiple pets kept in the huts.

The weight of carrying the monkeys, as well as other pets, creates a considerable impediment to a trekking lifestyle. One small monkey may have a negligible effect, but a single woman may try to carry three to four monkeys on her head as she leaves on a forest trek. The monkeys are difficult to manage, not only due to their raw weight, but also because they are moving, living beings that are interacting with each other as well as attempting to interact with the woman as she is trying to move quickly through the forest, perhaps with an infant of her own in her carrying strap. In essence, the practice can create a small community living on the woman as she attempts to go about her tasks. An early report of the Guajá, prior to FUNAI contact, described them as keeping large numbers of monkeys, particularly howler monkeys, as pets (Beghin 1957), indicating that the Guajá were trekking with their monkeys prior to their recent transition to a more settled lifestyle.

Older monkeys begin to display destructive and aggressive behaviors. While these are detrimental to the Guajá, and particularly Guajá children who are the usual targets of aggression, the greater detriment may be to the monkeys themselves.

THE CONSEQUENCES OF PET KEEPING
FOR THE MONKEYS

The natural behaviors of infant monkeys allow them to perform as children and substitute for children until maturational changes occur in juvenile monkeys creating more independent behavior, which makes it difficult for them to continue to fill the role. Negative consequences result for both the Guajá and the monkeys as aggressive behavior begins to emerge among the monkeys. The target of aggression tends to be the Guajá "siblings," the human children in the household. Aggression is best understood as abnormal behavior arising in monkeys as a result of their incorporation into the kinship system and its subsequent effects, including food provisioning and social isolation, which, when compounded with maturational changes, foster abnormal and aggressive behaviors.

Methodology

In order to evaluate the consequences of pet keeping for the monkeys, focal animal samples were conducted. Focal animal samples involve recording activity types, interactions with others, and the sequence and duration of behaviors (see Altmann 1974). In total, 61 focal animal samples were conducted on monkeys kept as pets in Guajá households representing approximately 130 hours (7,708 minutes). Conducting focal animal samples among the Guajá presented obstacles different from those found in either wild settings or in the zoo or laboratory setting. First, since the monkeys traveled with the Guajá women on treks and on day trips into the forest, only those monkeys that were available in and around the village were observed. In addition, each observation session usually lasted no longer than two hours because, after that period of time, I usually began to get indirect messages from household members that I was beginning to overstay my welcome. Making observations of monkeys inside households could also be difficult because of the activities of the household. Another problem I had not anticipated was the cockroach infestation of the houses. Some homes were infested so heavily that I had to swat the roaches off of me every minute or so as they dropped from the ceiling and crawled over me and often inside my clothing. I was never able to adjust fully to these conditions, and after two hour's time, I had usually reached the end of my tolerance as well.

Social Isolation

The nurturing and protective behavior displayed by the Guajá toward the pet monkeys is contingent on the ability of the monkeys to perform as dependent children. As the monkeys mature and become more dependent, Guajá behavior toward the monkeys changes. Older monkeys are often no longer considered *hamɨmɨrə*, and the Guajá reduce the amount of time spent in direct contact with the monkeys as they age. The focal animal samples demonstrated a correlation between the age category and the number of minutes spent in contact with the Guajá. A significant difference was found among the age categories of infant, juvenile, and adult ($\chi^2_{df = 2}$ = 2684.21, p < .001). Infants spent 79.77 percent of their time in contact with the Guajá, while juveniles spent 12.52 percent, and adults spent only 1.20 percent of the time in contact with the Guajá.

The changing role of the aging monkey in the Guajá household likely relates to the relatively short period of childhood dependency on parents displayed by Guajá children. As previously discussed, young girls marry and leave the natal household in early childhood. While boys remain in the natal household longer than girls do, they often leave for extended visits with various relatives away from their parents. Guajá child-rearing practices also encourage early independence and self-sufficiency. While the Guajá do provide a great deal of encouragement to their children, they rarely discourage or discipline behaviors, even when the behaviors may be dangerous. For example, toddlers were observed on several occasions to pick up knives and play with them. Rather than preventing the child from the behavior, parents try to redirect the child's attention to another activity. Thus, young children are neither viewed nor treated as completely dependent on the parents, and particularly on the care of the mother, once they are walking and weaned. For the monkeys, when they are past the clinging stage, they can no longer fill the role of dependent child. The behaviors of maturing monkeys become more and more of a nuisance to the Guajá, and they begin to tie them up for increasing amounts of time.

Lack of Social Contact with Conspecifics

Tying up the monkeys compounds the social isolation that begins when they are separated from members of their own species as infants. Older monkeys are typically tied up alone with little or no interaction with conspecifics or other monkeys species. Their interactions with the Guajá are often reduced to being handed or tossed foods and occasionally grooming Guajá heads for lice.

Activity budgets in the wild are available for two of the species the Guajá keep. They demonstrate significant differences from the Guajá's monkey companions, which would not be unexpected considering that the monkeys are kept in captivity as pets.[2] Bonvicino (1989) found that *Alouatta belzebul belzebul* spent, on average, 54.70 percent of the time resting, 21.15 percent travelling, and 7.20 percent eating and foraging, which is significantly different from the findings for the proportions of these major activities here with 84.98 percent resting, 2.80 percent travelling, and 9.89 percent eating ($\chi^2_{df=2}$ = 19.98, p < .001). Terborgh (1983) found that *Cebus apella* spent 12 percent of the time resting, 21 percent of the time travelling, and 66 percent of the time eating and foraging, which are also significantly different from the findings here with 29.60 percent resting, 6.87 percent travelling, and 9.35 percent eating ($\chi^2_{df=2}$ = 43.56, p < .001).

The activity budgets also reveal the character of the monkeys' social interactions. Social interaction accounted for 6.97 percent of the activity budgets for household monkeys. Total social interactions break down into 53.94 percent with the Guajá, 33.48 percent with a member of another monkey species, 10.99 percent with a conspecific, and 1.59 percent with another, non-monkey pet. Although these percentages demonstrate the general isolation from conspecifics, in that almost 90 percent of social interactions involve another species, this is largely due to Guajá behaviors, since most of the monkeys are tied up for much of the time and may not have the opportunity to interact with members of their own species.

However, a judgement can be made about the nature of the interaction with conspecifics and members of other primate species when such an opportunity is available. Significantly more affiliative behavior occurs with members of other species than with conspecifics ($\chi^2_{df=2}$ = 23.93, p < .001), with 57.30 percent of the behavior with conspecifics being affiliative, 65.07 percent of the behavior with other nonhuman primates being affiliative, and 82.75 percent of behaviors with humans being affiliative.

Emergence of Abnormal Behaviors

The activity budgets of the monkeys reveal a number of emerging abnormal behaviors that are coincident with the developmental changes that make them less like dependent children. First, as the monkeys begin to mature, they display a marked decrease in their clinging to a "mother's" head, and they begin to become more independent. Juvenile monkeys that are allowed to move freely begin to be a nuisance to the Guajá, as they race around the hut, get into the food supply, and pull apart the thatching of the walls and roof. This is particularly a problem with the capuchin monkeys, which engage in much manipulative activity, but it occurs to some extent in juveniles of all species. As a result, most of the monkeys are tied up for increasing amounts of time after they cease clinging to women's heads.

These older monkeys begin to exhibit a number of abnormal, non-productive, and sometimes self-abusive behaviors, such as pacing for hours on end, pulling out their hair, and biting their fingers and toes. Essentially, these are similar to the abnormal behaviors often observed in captive zoo or laboratory monkeys (e.g., de Waal 1989, Paterson 1992). In total, almost one-quarter (23.48 percent) of all activities included clearly disturbed behaviors. If resting is excluded, disturbed behaviors account for 42.23 percent of all active behaviors of the juvenile and adult monkeys.

As monkey aggression emerges in the juvenile monkeys, the target is usually another child in the Guajá household. Several injuries resulting from monkey aggression were observed. Brown capuchins caused severe injuries to several small children. One small boy suffered a deep bite to his buttocks, another boy had deep bleeding scratches to his feet, and a small girl received multiple gashes to her chest, fingers, and ears from brown capuchins in their respective households. One young boy had two large wads of hair yanked out by a Ka'apor capuchin. Two severe injuries by howlers were recorded. One boy was bitten several times on the head by a howler, and an adult woman had a deep bite to her leg from a howler. In addition, numerous minor bites and scratches occurred from monkey pets.

Despite the aggression of the monkeys towards the Guajá, they continue to keep them as pets and attempt to control them only by keeping them tied up. Monkeys are not the only pets that can cause injury. Serious injuries also occurred to two adult Guajá from a white-lipped peccary pet. The peccary had been acquired as an infant, but grew to adulthood among the Guajá and was kept in a pen of crosshatched wood beams. One evening, the peccary escaped and a husband and wife chased after it. Although they managed to retrieve it, both were deeply gashed on both the arms and legs by the peccary. Afterwards, there was talk about killing the peccary and eating it, but it could not be decided who would do it. After several weeks passed and the peccary still had not been killed, one man remorsefully stated that although white people knew how to kill their pets, they, the Guajá, could not do it. This phenomenon is related to the failed attempt by the FUNAI to get the Guajá to raise domesticated chickens. Once animals are associated with the household, the Guajá are reluctant to eat them, so the chickens are now kept as pets (although they eat their eggs).

The Effects of Food Provisioning

Since the pioneering work of Jane Goodall (1971) with Gombe chimpanzees, primatologists have been aware that food provisioning increases the level of aggressive behavior among nonhuman primates. Wheatley (1999:141) found

that tourists in Bali who fed long-tailed macaques (*Macaca fascicularis*) actually "trained" them to be aggressive because monkeys that attacked were ap-
peased and rewarded with food. Aggression among the Guajá's monkeys can
also be interpreted as, at least in part, related to their provisioning. One effect
of provisioning is to disrupt the normal activity levels of the monkeys as they
grow older. Provisioning the monkeys adds to the problem of the developmental increases in activity since the monkeys do not have the energy outlet
of foraging for foods. When they do "forage," they tend to do so inside the
Guajá huts and on the Guajá foods. These activity budgets are dissimilar from
what is observed in monkeys in the wild. In specific reference to eating and
foraging, Terborgh (1983) found that *Cebus apella* spent 66 percent of its time
eating and foraging and *Saimiri sciureus* spent 61 percent of its time foraging,
both of which are more than three times the amount of time spent by these
species among the Guajá.

The nutrition of the monkeys is also affected by provisioning. Despite the
Guajá's extensive knowledge of the natural diets of the monkeys (as addressed in
chapter three), the monkeys are typically fed the same food items as the Guajá.
Occasionally, foods are procured for the monkeys that the Guajá do not eat, but
typically no special provisions are made, and the monkeys eat manioc, squash,
and meat, sometimes eating members of their own species. All species, except
the squirrel monkey, were observed eating howler meat, and both species of capuchin were observed eating meat from their own species that was given to
them by the Guajá. The Guajá state that the monkeys are eating their siblings,
harɘpihárɘ. Monkeys are basically fed the same food items of the Guajá and eat
when the Guajá eat. Infant monkeys are breast-fed, and it is not unusual to see
a woman with a monkey on one breast and her own infant on the other. As
young juveniles, squash and manioc are introduced, and, later, game meat is
fed to the monkeys, including monkey meat. Women premasticate foods and
feed them to the monkeys, with the monkeys often eating directly from a
woman's mouth. One method of quieting a monkey that is upset is to have it
drink saliva from a human mouth.

One interesting observation was that several brown capuchin monkeys were
observed cracking open long bones and eating the softer, spongy bone inside.
Two methods for accomplishing this were observed. One male monkey was observed on several occasions laying a bone along a horizontal house post and
then pounding on the bone with a babassu nut. Other monkeys were observed
chewing off the ends of bones to reach the inner spongy bone and marrow. Another unusual diet item observed was tortoise carapaces for the black-bearded
sakis. In one household that had two sakis, tortoise carapaces were hung near
the monkeys, and they chewed off pieces and consumed them. In addition,

members of the species *Cebus kaapori* and *Chiropotes satanas* were observed eating large quantities of the cockroaches that infested the Guajá huts, which may have adversely affected their health through the spread of disease.

Eating-time budgets from the focal animal samples found that 41.60 percent of the diet of the monkeys overall derives from domesticated plants, with an additional 15.09 percent coming from fish and game acquired by the Guajá. Thus, the bulk of their foods consists of provisioned foods that they either do not have access to or would have limited access to in the forest (although crop-raiding by some species may introduce domesticates into their diet). Since the Guajá have only recently adopted domesticated plants, it is possible that in the past the provisioned diets of the monkeys more closely resembled natural diets. For the howlers, any provisioned diet may have been difficult for the Guajá to replicate since howlers rely heavily on flowers and leaves, which are not part of the Guajá diet and may be hard to reach in the canopy. In the Atlantic forest, the diet of the related subspecies *Alouatta belzebul belzebul* was found to consist of 59.0 percent fruit, 27.6 percent flowers, and 13.3 percent leaves (Bonvicino 1989:170), which is significantly different from the diet of the howlers among the Guajá, which included no flowers or leaves ($x^2_{df = 2}$ = 50.57, p < .001). The diet of the black-bearded sakis likely was not replicated by the Guajá in the past. In Surinam, the related subspecies *Chiropotes satanas chiropotes* spent 62.2 percent of the time feeding on immature seeds, 30.0 percent of the time feeding on ripe fruit, and 3.4 percent of the time feeding on leaf stalks and flowers (Mittermeier and van Roosmalen 1981; Mittermeier, et al. 1983; van Roosmalen, Mittermeier, and Fleagle 1988). The diet of *Chiropotes satanas* among the Guajá was approximately 48.50 percent animal material, and they were not observed to eat seeds, leaves, or flowers. If comparing the relative amount of plant and animal material in the diet between black-bearded sakis among the Guajá and those in Surinam, a significant difference exists ($x^2_{df = 1}$ = 64.1, p < .001). The differential survival rate of the capuchins (to be discussed below) may stem in large part from their relative omnivory and ability to survive on Guajá foods.

Life Expectancy

The Guajá kept a total of ninety monkeys between February of 1996 and August of 1997. The vast majority (92.22 percent) were infants and juveniles. The relative frequency of monkey species kept by the Guajá can be evaluated along two axes: the overall relative frequencies of the monkey species during the study period and the relative frequency of monkey species in terms of new acquisitions.

While *Cebus apella* is the most abundant monkey pet, slightly more *Alouatta belzebul* monkeys were introduced to the group, which reflects the differential attrition of *Cebus apella* and *Alouatta belzebul* among the Guajá. The order of the frequency of occurrence of monkey species in the community was: *Cebus apella* (36.67 percent), *Alouatta belzebul* (25.56 percent), *Cebus kaapori* (16.67 percent), *Chiropotes satanas* (7.78 percent), *Saguinus midas* (7.78 percent), *Saimiri sciureus* (3.33 percent) , and *Aotus infulatus* (2.22 percent).

The attrition rates of primates differed by species. Here, attrition indicates a loss from the community through either death or escape. In terms of new additions during the study period, 39.22 percent (N = 20) were *Alouatta belzebul*, 37.26 percent (N = 19) were *Cebus apella*, 15.69 percent (N = 8) were *Cebus kaapori*, 5.88 percent (N = 3) were *Saguinus midas*, 3.92 percent (N = 2) were *Chiropotes satanas*, and 1.96 percent (N = 1) were *Saimiri sciureus*. The two most frequently occurring species kept as pets were *Alouatta belzebul* and *Cebus apella*. In comparing these two species, a significant difference ($\chi^2_{df = 1}$ = 19.96, <.001) was found in their attrition. Overall, 78.26 percent of the howlers were lost from the group while only 18.18 percent of the *Cebus apella* were lost.

One way to determine the significance of the attrition rate of Guajá monkey pets is to make a comparison with New World monkeys in the wild. Demographic data are available for *Alouatta palliata* howling monkeys in Costa Rica, whose one-year infant survival rate is 55 percent (Clarke and Glander 1984:115). A category roughly comparable to the infant survival rate was created to track "new addition" survivorship among the Guajá's pet monkeys. New additions include only those pets acquired during the first six months of the research period; at least a full year of monitoring is available for each of them. The new additions were either infants or very young juveniles and thus of a similar age grade to the infant howlers tracked by Clarke and Glander (1984). Overall, the one-year survival rate for Guajá pets is only 44.83 percent, compared to 55 percent for infant howler monkeys in the wild. The 44.83 percent survivorship for Guajá pets may even be an overestimate, since runaways whose fate was unknown were not included; likely, many did not survive.

The one year survival rate of 44.83 percent does not reflect the variation among the species, which is considerable. Of the new additions in the first six months of the research period, only the capuchins had survivors twelve months later, with the brown capuchin having a 75.00 percent twelve-month survival rate and the Ka'apor capuchin having a 57.14 percent twelve-month survival rate. However, an adult howler monkey, an owl monkey, and a tamarin were present in the group. The indication is that these species rarely survive to adulthood among the Guajá.

Possible Benefits of Captivity

In general, the consequences for the monkeys in Guajá captivity are largely detrimental. However, many of those infants whose mothers abandoned them during the hunt or were killed would likely die if not taken in by the Guajá. Thus, the Guajá do provide at least a temporary extension of life for the infants. In general, the mortality rate of the monkeys is higher than would be expected in the wild, but species-specific differences should also be considered. Although monkeys of all species rarely survive to adulthood among the Guajá, the capuchins do have a high rate of one-year survivorship, possibly due in part to predator protection offered by the Guajá. However, predator protection is not absolute, since one woman's pet tamarin was snatched by a harpy eagle during the study period. Chances of long-term survival in captivity with the Guajá are not good for any of the monkey species. It should be mentioned, though, that two brown capuchins have been known to reproduce among the Guajá. One capuchin birth occurred during the study period in 1997, and Balée (personal communication) noted the birth of another capuchin in the village in 1995.

In addition, two howlers are known to have returned to the wild and survived. The monkeys would return periodically to the vicinity of the Guajá settlement where their presence was recognized by their territorial calls. The Guajá provisioned them when they were in the area but no longer had physical contact with them. In fact, one individual was bitten by one of the monkeys when she accidentally came upon it while it was in the vicinity. These monkeys are not hunted and are still recognized as belonging to their original owners. The female did give birth to an infant. It is possible that members of other species have returned successfully to the wild, although their presence is not so readily determined as is that of the howlers with their loud territorial calls. A similar phenomenon of pet monkeys returning to the forest has been reported by Shepard (1997) among the Matsigenka, who also raise monkeys as pets. The former pets living in the wild are not eaten and are seen as having a practical value as living decoys to attract other wild monkey troops.

Pet Keeping, Semi-Domestication, and Domestication

Queiroz and Kipnis (1991) suggested that the monkeys kept by the Guajá were "semi-domesticated" since they are kept as pets and because some of them (*Cebus* spp.) were observed to attack strangers, similar to guard dog behavior.[3]

The research here suggests that monkey aggression can be characterized as abnormal behavior deriving from social deprivation and other stressors resulting from having been raised in captivity by the Guajá. For my part, as an "intruder" I did not experience any physical aggression from the monkeys, with the exception of one tamarin, even during my initial encounters with them when I first began the research. In addition, monkeys were not observed acting aggressively towards the other strangers who visited the Guajá area. However, the Guajá clearly identified some of the monkeys as prone to biting. Once such behavior emerged, the Guajá began to tie up the offending monkey to prevent it from continuing. In this light, many of the monkeys cannot even be considered tamed, much less domesticated. The monkey aggression was global rather than directed specifically at an intruder. Children suffered the most from the aggression because they either attempted to play with aggressive monkeys or unwittingly moved within striking distance of an aggressive monkey. The tying up of aggressive monkeys limited their access to intruders, but not to children who are frequently nearby in the huts.

The tamarin that acted aggressively towards me did so after I had been in the field for over a year. I had been talking with a woman when her tamarin attempted to jump from her head to mine. It startled me and I jumped back and the tamarin landed on the ground and bit me on the leg. The next day, as I left the same woman's house during my daily rounds, the tamarin jumped from the roof, pounced on my head, and then ran away. Although the behavior was aggressive, the tamarin's actions seemed to be based on her familiarity with me rather than her perceiving me as a stranger. Further, the ambush clearly had no protective role, since it occurred while I was leaving the hut instead of when I entered it. In contrast, the dogs kept by the Guajá did display direct aggression towards intruders and did guard the village. Throughout my fieldwork, the dogs behaved with extreme aggression. At least once a week, they would form a pack around me, barking, snarling, and snapping. I never went into the village without babassu nuts to throw at them and a big stick, which I sometimes had to use to fight them off. The dogs displayed the same behavior to all strangers coming into the Guajá village.

The alarm calls of monkeys may possibly serve as warnings to the Guajá of impending intruders. However, the Guajá themselves have a remarkable sensitivity to the sounds of their environment. On many occasions, the Guajá were observed to know when someone was approaching the village well before he or she was in view (and my personal earshot) and also able to identify the specific individual who was approaching. Although monkey warning calls should not be entirely discounted as useful to the Guajá, as hunter-gatherers who spend their lives within the forest, such warnings may not be necessary for them since their abilities probably at least rival, and perhaps even exceed, that of the monkeys.

However, the more general question remains as to whether pet keeping by the Guajá can be considered domestication or semi-domestication in some form. It has been hypothesized, although not yet demonstrated, that animal domestication first originated with pet keeping (Zeuner 1963). The domestication of an animal, as a process, involves influencing reproductive behavior and, thus, gene frequencies (Bökönyi 1969, Clutton-Brock 1989, Ingold 1980). Although two capuchins did reproduce in captivity, this should be characterized as a rare occurrence. At this point, there is insufficient evidence to support the idea that the Guajá influence the gene frequencies of the monkeys.

However, the notion of semi-domestication remains intriguing in that the traits valued by the Guajá in the monkeys are neotenous. One of the traits used to distinguish domesticated from non-domesticated animals is neoteny, or the retention of juvenile features (e.g., Zeuner 1963). The Guajá demonstrate a clear preference for juvenile traits in monkeys, and young monkeys clearly receive better care from the Guajá. Since monkeys have reproduced in Guajá captivity and behavioral neoteny could provide a survival advantage among the Guajá, it is not outside the realm of possibility that under such conditions, artificial selection could occur over time for neotenous monkeys. The evidence here suggests that this has not yet occurred, and the majority of monkeys (99 percent in the study period), were replacements from the forest rather than bred in captivity.

However, it should be mentioned that Sponsel, Natadecha-Sponsel, and Ruttanadakul (in press) have challenged the suitability of the categories of domesticate and nondomesticate for macaques in Thailand. They argue that although there is no artificial selection, the macaques can be considered semidomesticates due to the Thai practices of naming them, providing shelter, provisioning, bathing, grooming, burying at death, and using their labor to harvest coconuts (Sponsel, Natadecha-Sponsel, and Ruttanadakul, in press). If their criteria are used, it would seem that monkeys kept by the Guajá could be considered semidomesticates as well.

In conclusion, the role of the monkey in the Guajá social system is predicated on the natural social behavior of monkeys. Infant monkey behavior is similar enough to human infant behavior that infant monkeys can substitute as surrogates and serve as a means to enhance the image of Guajá female fertility. However, it is impossible for the monkeys to sustain this role due to the compounding effects of maturational changes, food provisioning, and social isolation; the result is older monkeys who display aberrant and often aggressive behaviors if they survive at all. In the following chapter, the importance of the sibling relationship among the Guajá will be demonstrated to extend beyond the kinship system and will be seen as integral to understanding how the Guajá interpret their relationships to the natural world and the spiritual world, with monkeys having a central role in mediating between nature and culture.

7

COSMOLOGY AND SYMBOLIC CANNIBALISM

The notion of reflexivity has been used as a means of interpreting the process of projecting one's own cultural values onto others. Ohnuki-Tierney (1984) defines reflexivity as the sense of distance from the self to allow the self to become an object of study.[1] In cultural anthropology, it suggests that the interpretation of other cultures may reveal more about the culture of the observer than that of the observed. Within a group, animals can serve as reflexive symbols. For example, Geertz (1973) described cocks in Bali as symbolic expressions or magnifications of the self that also delineate the boundary between humanity and animality. Animals that have a material importance to a group often have symbolic importance as well. Among some African pastoralists, such as the Nuer and the Tshidi Barolong, cattle are of key material importance as well as being key in the construction of individual identity and collective group identity (Comaroff and Comaroff 1990; Evans-Pritchard 1940).

Nonhuman primates may provide a cultural Rorschach for many human groups. The close physical and behavioral similarities of other primate species to human beings makes them particularly amenable to the projections of cultural values. As Haraway (1989) has described, even primatologists have not been entirely immune to the projection of their biases in interpreting nonhuman primate social behavior. Ohnuki-Tierney uses reflexivity as a means to understand the historical meanings of the monkey as a metaphor for the self

FIGURE 7.1   The owl monkey

among the Japanese (Ohnuki-Tierney 1984, 1987, 1990; also see Wolfe and Gray 1992). Burton (2002) provides a similar analysis of the historical meanings of the Monkey King in China.

NONHUMAN-PRIMATE SYMBOLISM
IN A CROSS-CULTURAL PERSPECTIVE

Identification with nonhuman primates is often revealed in dietary practices where they may be alternately avoided or preferred as food due to their association with human beings. Several lowland South American cultures indicate a preference for monkeys due to their physical similarity to human beings. The Suyá highly value certain species of monkey as food sources, and monkeys are also important figures in their dream symbolism and in ceremonies due to their physical similarity to humans and their calls (Seeger 1981). Among the Huaorani, monkeys account for 36 percent (by weight) of the game (Yost and Kelley 1983), and Huaorani women also breast-feed pet infant monkeys, considering them kin, and even give them ritual burials (Rival 1993). The Kalapalo of the Upper Xingu consider land animals disgusting to eat, with the exception of monkeys (and sometimes coatis), which are classified as being similar to humans and are central in their *kaiju* (capuchin monkey) redistribution ceremony (Basso 1973).

Among the Barí of Venezuela, monkeys are heavily exploited as food and kept as pets, but they identify differently with howlers and spider monkeys (Lizarralde 1997, 2002). Spider monkeys are the preferred game, the most frequently hunted, and the most commonly kept pet. The Barí believe they were created from spider monkeys and that wearing their teeth confers manual dexterity. Howlers, however, are not kept as pets and are believed to be of low intelligence and slow speed. While spider monkey teeth are valued for necklaces, howler teeth are not. The Barí have also viewed the ability to eat monkeys as representing the boundary between the Barí and non-indigenous people; however, this is now changing. Westernized Barí tend to avoid eating monkeys because of their similarity to humans and because they are teased by non-indigenous people as "monkey-eaters."

Crocker (1985) suggests a kind of magical contagion among the Bororo, who eat monkeys that are considered to epitomize speed and grace. Monkeys figure prominently in Bororo spiritual life and are associated with *bope*, which Crocker describes as both a principle of organic transformation and as entities. Monkeys in general are frequent animal familiars for *bope*, including the *juko*, which are a class of *bope* that look like capuchins. The process of becoming a shaman involves being surprised and spoken to by a howler. If a *bope* resembling a howler is seen at other times, it presages the death of the viewer or a kinsperson.

In some cultures, both within and outside of Amazonia, monkeys are avoided as food because of their similarity to human beings. Among the Kagwahív (Parintintin), monkeys are kept as pets, but avoided as food due to their similarity to human beings (Kracke 1978). Mittermeier and Coimbra-Filho (1977) described the disgust of some non-indigenous Brazilian peoples at eating *Cacajao* monkeys because of the human appearance of its bald red face. It has been argued that the Islamic prohibition on eating monkeys is related to their physical similarities to human beings, which would make such consumption "close to cannibalism" (see Sponsel, Natadecha-Sponsel, and Ruttanadakul, in press, MS. 5). In Africa, the Ju/'hoansi (!Kung-San Bushmen) of the Kalahari Desert eat virtually all the animals in their habitat with the exception of the baboon and the hyena. The baboon is not eaten, "on account of its being so like a man," and the hyena is not eaten because, "it eats human corpses" (Schapera 1951: 93). Among the Fang of the Republic of Equatorial Guinea, chimpanzees are not eaten because it is considered sinful to consume something nearly human (Sabater Pí and Groves 1972). Sabater Pí and Groves suspect that this belief is due to missionary influence, since the Fang reportedly practiced cannibalism until the turn of the century. Further, they report that gorillas are eaten, and that the most preferred parts are the facial muscles, palms, soles, and tongue, also the most preferred human parts eaten in their cannibalism. Similarly, Sicotte and Uwngeli (2002) suggest that the reason the Hutu have traditionally avoided eating mountain gorillas is that it would seem cannibalistic.

In several cultures, nonhuman primates are believed to share a common origin with human beings. Several lowland South American cultures that eat monkeys have this belief, such as the Barí (Lizarralde 1997, 2002), the Matsigenka (Shepard 2002), the Mehinaku (Gregor 1977), and the Guajá. The Wari' believe that spider monkeys have partially human origins, being descended from the union of a Wari' woman and a male spider monkey (all the originary spider monkeys in their belief were male) (Conklin 2001). In Africa, the Mende of Sierra Leone believe that chimpanzees and human beings have a common origin. Richards (1993:145) goes as far as to describe the Mende as primatologists because they identify with chimpanzees due to similarities observed in chimpanzee natural behavior, including tool use and medicinal plant use.

Reference to monkeys can also be a commentary on the human status of a people. In some groups, calling another human group "monkey" diminishes their humanity. For example, the Tupían term *kaya-po* refers to people who resemble monkeys (Werner 1984:173). In other cases, monkeys are considered to be more closely related to a people than are other human groups. The Mehinaku do not consider non–Xingu Indians to be fully human, and they are called by the negative term *wajaiyu*. Monkeys are classified as human beings (*neunéi*), a group including Xinguanos, Brazilians, and other Westerners (Gregor 1977). Among the Huaorani, many myths involve social catastrophes caused by monkeys who overstep their bounds in either trying to be too close to human beings or too distant from human beings (Rival 1996). Similarly, Zulu mythology involves themes of baboons desiring to capture and raise a human child and stories of baboons disguising themselves as humans (Carter and Carter 1999). Primates can also play a role in ethnic politics. According to Sicotte and Uwengeli (2002), Hutu informants in Rwanda viewed mountain gorillas as possessing more desirable human qualities than the Twa people.

Such formulations are also found in Western constructions of delineating the self/other boundary. In early-nineteenth-century Brazil, one expression of the tension between the Portuguese and native born Brazilians was the description of the white faced capuchin as "macaco português" which changed by the end of the century to "macaco inglês" (Cleary 1998:131). Another example comes from the nineteenth-century Amazon explorer Louis Agassiz, who compared Africans with Amerindians in saying, "I may say that if the negro by his bearing recalls the slender, active Hylobates, the Indian is more like the slow, inactive, stout orang" (Agassiz and Agassiz 1868:530). Wheatley (1999:18–21) has described a number of such examples of apes delineating the line between self and other. For example, he notes that Comte de Buffon wrote in 1871 that the apes and Hottentots should be classified together, marking them both as nonhuman. Similarly, he writes that nineteenth-century English cartoonists suggested a similar designation for the Irish by depicting them as apes. He also notes that the term "ape" was applied by early Christians to non-Christians.

According to Carter and Carter (1999), the image of the ape became associated with the image of Satan in Medieval Christianity because, as Satan attempted to imitate God, the apes were imitators of humans. The simian as a religiously degraded human is also found in Islamic traditions. Referring to someone as a monkey is one of the most degrading insults one can give in Islam, and the belief also exists that God may punish humans by turning them into monkeys (Sponsel, Natadecha-Sponsel, and Ruttanadakul, in press). Comparing another to a monkey or ape is also a common epithet in the English language, such as was seen in Karl Marx's expressing contempt for the work of Thomas Malthus by calling him a baboon (Harris 1968).

Monkeys may also take the role of the clown in some cultures, parodying humans and providing a means for social criticism. Among the Mekranoti, a ritual clown's role is inherited by a man from his mother's brother and requires the man to wear a howler mask during ceremonies, making fun of everything around him (Werner 1984:175). Carter and Carter (1999:275) also note that in a number of American movies, nonhuman primates imitate humans and provide a means of comic relief, such as Tarzan's chimpanzee companion Cheetah and Clint Eastwood's orangutan sidekick in *Every Which Way But Loose*. More generally, monkeys and apes are used in American television and entertainment as commentary on human life. In the "monkey tea party" in its varied forms, nonhuman primates dress as humans and reenact complex norms, or alternately, they flaunt norms, and producers show them going "wild" with credit cards or in sports cars.

In several Old World cultures, monkeys are deified or considered sacred. The baboon was deified by the ancient Egyptians and is depicted in Egyptian carvings and hieroglyphics performing various tasks related to the sun god (see Kalter 1977). Buddhists believe that one of the incarnations of the Buddha was a monkey (Sponsel, Natadecha-Sponsel, and Ruttanadakul, in press; Sponsel, Ruttanadakul, and P. Natadecha-Sponsel 2002). In cultures influenced by Hindu religion, monkeys are important because of their association with the monkey god Hanuman and the Hindu epic *Ramayana*, which dates to about 2,500 B.C. In the epic, Hanuman, a faithful monkey servant, helps Prince Rama rescue his wife Sita from the evil Rawana (see e.g., Oppenheimer 1977; Small 1994; Wheatley 1999; Wolfe 1992, 2002). In India, Hanuman langurs (*Presbytis entellus*) are protected and provisioned in the cities because of both their association with Hanuman and Buddhist religious beliefs which forbid the killing of animals (Oppenheimer 1977). In North India, Rhesus macaques (*Macaca mulatta*) are fed in reverence to Hanuman, sometimes as a magical means to prosperity or in association with a special occasion such as a wedding, birth, or festival (Wolfe 1992). Toque macaques (*Macaca sinica*) receive similar treatment in Sri Lanka due to their association with Hanuman (Dittus 1977). Monkeys have an important religious role in the culture of the Balinese people, which has been described in Wheatley's (1999) recent book. The name of the Indonesian island itself is probably derived from a mythical monkey. The monkey priest "Bali" (or Subali) is the brother of Sugriwa, who helps Hanuman rescue Sita in the *Ramayana* epic (Wheatley 1999:10–11).

CANNIBALISM

Animistic beliefs among many groups both within and outside of Amazonia can be considered inherently cannibalistic if kinship or personhood is attributed to the forms of life that are preyed upon for food. Cannibalism among the Amazonian groups has been a subject of fascination for the Western world at least since the time of Michel de Montaigne's sixteenth-century essay "Of Canni-bals," describing Tupinambá cannibalism (Goldman 1999). In Arens's (1979) now classic work, he cast doubt on the idea that cannibalism ever actually ex-isted in any culture. He argues that the idea of cannibalism is universal insofar as human groups attribute it others, but he was not able to uncover adequate documentation of cannibalism as an actual custom in any of the societies he re-viewed where it had been described. In the eighteenth-century "tropa de res-gates," the Portuguese used accusations of cannibalism to justify enslavement of Amerindians (Sweet 1975:116–17). Here, slavery was rationalized as a means of rescuing Amerindians from being eaten. Since Arens's work, there has been credible ethnographic evidence of cannibalistic practices, but his argument is certainly important, in that ethnocentric demonization of others as cannibals is insufficient evidence of the actual practice.

A more recent example of the phenomenon Arens describes are the national headlines that were made by the suggestion that the Anasazi of the American Southwest may have been cannibals. However, Kantner (1999) warns that while it is clear that the Anasazi practiced violent mutilation of bodies, it cannot be as-sumed that this meant cannibalism. Similarly, Pickering (1999) challenges the evidence that cannibalism was practiced by some Aboriginal Australian groups. Here, mortuary dismemberment rites have been confused with cannibalism or human sacrifice. Among the Ku Waru of Highland New Guinea, the word for cannibal and the word for foreigner are almost synonymous, and they attributed cannibalistic practices to the Euro-Australians they first came into contact with in the 1930s (Rumsey 1999:107).

Apart from what Goldman (1999) describes as survival cannibalism—aber-rant behaviors in emergency situations—there is credible ethnographic evi-dence of the practice among some human groups. For example, Gardner (1999) believes there is clear evidence that the Mianmin of New Guinea practiced cannibalism. The Onabasulu of New Guinea have been described as eating those accused of witchcraft (Ernst 1999). In Amazonia, two recent examples are the work of Conklin (2001) on the Wari' mortuary cannibalism and Whitehead (2001) on cannibalism associated with the Kanaìma death cult in Guyana.

GUAJÁ SYMBOLIC CANNIBALISM

Anthropophagy has been practiced in some Amazonian groups, but, more gen-erally, cannibalism is a common trope for understanding death and relations of consumption. Given that Guajá animistic beliefs involve social relations with nonhuman life forms, consuming them can be considered a form of symbolic cannibalism. This is particularly true in the case of howlers, who are considered to be the form of life most closely related to the Guajá.

The Guajá deny ever having literally practiced cannibalism, although one older Guajá man reports that approximately thirty years ago a major conflict occurred between the Guajá and the *kamara* (non-Guajá Indians) wherein many on both sides were killed and that the *kamara* ate the dead Guajá. One man claims that his mother was eaten by the *kamara* when he was approximately six or seven years old. It is possible that the *kamara* described were one of the Tupí or non-Tupí speakers in the area. However, it is also very possible that this is an example of demonizing enemies through attributing cannibalistic practices to them.

The Guajá engage in a form of symbolic cannibalism based on two principles. The first is the general level of meta-affinal sociality that the Guajá employ in describing their relationships to nonhuman beings. If forms of life that they prey upon are considered kin, then consumption of such kin can be considered a form of symbolic cannibalism. Secondly, in the Guajá view, partial consanguinity among life forms leads them to specialize in hunting prey to which they are most closely related. The principle does not employ neat cycles of energy exchange, but rather represents a logical association of preferential consumption of partial consanguines which has both predatory and reciprocal elements and is reflected in the Guajá creation myth, the activities of cannibal ghosts, and in the *harapihára* siblingship associations of the divinities with plant and animal spirits.

Creation Myth

Many Amazonian groups locate animal origins in humanity rather than human origins in animals (see Viveiros de Castro 1999b). Most of the Guajá game animals have human origins, although some involve transformations from plants or landscape features. The creation of human beings is told as follows: The creator divinity, Mai'ira had a large erection and a strong sexual desire and came upon a tree with large breasts and female genitalia who is called Yapo-kwĩya (also called Yapo-ira or Yapiawɨ-wɨra). He cut the tree down to human size and ate *inaya* palm fruit (*Maximiliana maripa*) (or in some versions, the *kamičia* tortoise [*Geocholone* spp.]) through both his penis and mouth. He then had sex with the tree and blew on her or sang to her or both to make her human. She became pregnant with twin boys who are called Mai'ira (the same name as his father, or in some versions *imana*) and Mukúra-či'a. One day, Mai'ira (the father) went walking in the forest in order to find some wood to make a bow. His wife (the former tree) decided she wanted to follow after him, but she did not know the forest. The still-fetal male twins told her the way and she set off after him. A large jaguar called Yawárə-awa (jaguar man), who was described as having an engorged stomach and the anthropomorphic features of bipedalism and lacking fur, came upon them and devoured Yapo-kwĩya but not her twins. When Mai'ira discovered what had happened, he was very afraid of the jaguar and left the earth forever and went to the *iwa*. The jaguar took the twins and raised them in the forest, where they learned to eat the traditional foods of the Guajá, including how to crack open the *wã'i* palm fruit (*Orbignya phalerata*), cook the *muki'a* fruit (*Caryocar villosum*), beat *inaya* fruit into a liquid soup, and how to hunt and eat various game animals.

The creation myth seems to encode the bifurcation of the Guajá soul at death. As previously mentioned, the Guajá metaphorically link the opossum, *mukúrə*, with the *aiyã*. Thus, one twin, Mukúrə-či'a, seems to represent the part of the soul that remains on earth as a cannibal ghost. The other twin is named after the creator, Mai'ira, and seems to represent the sacred form of the soul destined for the *iwa*.

The creator, Mai'ira, is also responsible for making almost all the other forms of life in the universe. The howlers (*wari*) were originally created from human beings. Mai'ira came upon a group of Guajá who were high in a tree eating *aparihu* (*Manilkara huberi*) (*čičipea* [*Inga* spp.] in some versions). Mai'ira clapped his hands or stomped his feet or blew on them, or any combination of those, and those Guajá were transformed into howlers who then ran away. Later, other Guajá saw the howlers and thought they were *awa* (people) , but Mai'ira told them they were *wari* and that they should eat them. In some versions, the brown capuchin, Ka'apor capuchin, and squirrel monkeys were created from the Guajá as well.

The transformation of the Guajá into howlers does not seem to involve any specific judgment or punishment by Mai'ira, but it seems that the behavior of eating the *čičipea* or *aparihu* fruit in a tree is what prompted Mai'ira to change them into human beings. With the exception of children eating certain *Inga* species, the Guajá say that they do not eat *aparihu* or *čičipea* and that those fruits are eaten by monkeys. Some informants seemed to suggest that if they were to eat *čičipea* or *aparihu*, they might turn into monkeys.

The story of the creation of howlers from the Guajá was consistent across informants. However, the story of how other monkeys were created varied. Some informants described other monkey species as being created directly from human beings or from the transformations of other substances, but more commonly their origins were in the *aiyã* cannibal ghosts. In the first case, it is possible that the howler creation myth is being generalized to other monkeys and that the howler serves as a monkey prototype (see Rosch 1978). The general version of the creation of other monkey species from the *aiyā* is told through a story of a battle that once occurred wherein many Guajá were killed, generating many *aiyā*. Mai'ira hit the *aiyā* with his bow, and the brown capuchin, Ka'apor capuchin, the owl monkey, the bearded saki, the squirrel monkey, and the tamarin ran out. Other game species were also created from the *aiyā*, including the agouti, paca, rabbit, deer, coatimundi, armadillo, and opossum.

Several species have their origins in plant foods that they consume. An additional origin myth for the brown capuchin tells of its being transformed from a *wā'ɨ* palm nut (*Orbignya phalerata*) after Mai'ira shot an arrow into the palm. The Guajá report that the brown capuchin eats the palm in the wild. The squirrel monkey is said to be made from the *mariawa* palm (*Bactris* sp.), which it eats. The agouti and paca eat a number of palms fruits and are said to be made, respectively, from the *pinōwa* (*Oenocarpus distichus*) palm and the *yahárɔ* (*Euterpe oleracea*) palm. Bearded sakis are also said to be created from *kipi'i*, which is a kind of termite nest, or from plant leaves, *ka'aro*. While bearded sakis do eat insects, the association is more related to the similarity in their appearance to the large, dark termite mounds. Two other monkeys were created from plants, but the Guajá did not report these monkeys as eating the plants: the owl monkey from the leaves of the *akarahu* plant (*Calathea roses-picta*), and the tamarins are also said to be made from leaves of the *yahárɔ* palm (*Euterpe oleracea*) or the *wā'ɨ* palm.

Some of the other game major game species have alternate origin myths other than creation from the *aiyā*. Deer are also said to be made from *tami'ki'i*, which is another type of termite mound. The coatimundi is also said to have been made from an opossum who was eating *inaya* when Mai'ira blew on it and changed it into a coatimundi. The peccaries and tapirs were created through Mai'ira either whistling or blowing on stones. Some say that the peccaries were made from *kipi'i* (termite mounds). Some also say that the tapir was made specifically for the jaguar after he asked Mai'ira for one. Fish, birds, and insects were made by Mai'ira blowing on leaves. There are several versions of the jaguar's creation. In one, Mai'ira simply blew on wood and created it. In another, the jaguar was made from Mai'ira blowing on the *yawaka'a* (jaguar-herb) plant (*Psychotria* sp.). In another version, a tiny jaguar sprang from Mai'ira's head, who then shot the jaguar, causing it to become giant. Then, as the jaguar ran either through or near the river, piranhas jumped up and tore at the jaguar, causing it to divide into many smaller jaguars.

The origin of the *kamičia*, the red-footed and yellow-footed tortoises, is somewhat different. Some say that Mai'ira did not make the *kamičia* at all, but that it was made by the divinity Kiripi. Mai'ira later saw the *kamičia* running by and captured it and broke its legs, slowing it down so that it would be easier to catch and eat. Others say that Mai'ira cut wood and had sex with it, and that the

kamičia came from his semen. He then ate the *kamičia* through both his mouth
and penis. The latter version is similar to the story of the creation of the first
woman and probably relates to the importance of the *kamičia* to fertility.

The specific origins are known for only a few of the plants. The *wā'ɨ, inaya,* and
pinõwa were made after the Guajá complained to Mai'ira that they had nothing to
eat. He then blew and brought these palm fruit trees into existence. In another version, Mai'ira made the *wā'ɨ* palm nut from a stone. The *inaya* palm was also said to
be made from the semen of Mai'ira. The *irapárə* tree (*Tabebuia* sp.), the source of
bow wood, was made from snakes. Interestingly, the Guajá say that Mai'ira did not
make snakes at all, and they do not seem to have a specific origin myth.

Cannibalistic themes emerging from these myths comprise the predation of
the Guajá on former Guajá (howlers), on twice transformed Guajá (numerous
game animals), and the transformation of an animal from a plant that it eats
(e.g., the brown capuchin and the squirrel monkey). In terms of the twice transformed Guajá, at first glance it might appear that a cycle exists here with the
aiyā cannibalizing the Guajá and the Guajá eating transformed versions of the
aiyā. However, the transformation of animals from the *aiyā* was a one-time
event bringing these animals into being, and they are no longer created from
aiyā. The *aiyā* themselves can only be killed by jaguars.

Karawa Divinities and Spirit Harəpihárə

A number of the *karawa* divinities are associated with a particular animal over
which the divinity has principal control. For example, the divinity Yu-či'a hunts
howlers in the *iwa* and has the ability to bring howlers near the Guajá so that
they can hunt them. Some of the divinities control more than one animal, but
these animals tend to be closely related in either appearance or behavior. For
example, the divinity Apario'ɨ has control over both the owl monkey and the
kinkajou, which are both nocturnal, arboreal species.

The *karawa* divinities also have a *haĩma* (spiritual) *harəpihárə* sibling from
which they draw their names. Often, but not always, there is a relationship between the spirit *harəpihárə* of the divinity and the animal species that the divinity controls. For example, the squirrel monkey is controlled by the divinity
called Mariawa'ɨ. Mariawa'ɨ is the spirit *harəpihárə* of the palm *mariawa'ɨ*,
which is eaten by the squirrel monkey. In addition, as noted above, in some versions of the creation myth the squirrel monkey was also created from the *mariawa'ɨ* palm. Table 7.1 is a summary of the animal species, what each was created
from, its associated divinity, and whether any cannibalistic theme is operating.

TABLE 7.1 Cannibalistic Associations

Guajá Name	Common Name	Created From	Divinity	Endocannibalism
awa	Guajá	Mai'ira and tree	Mai'ira	*aiyã* are former Guajá and eat Guajá
aiyã	"cannibal ghosts"	Guajá		*aiyã* are former Guajá and eat Guajá
wari	howler monkey	Guajá	Yu-či'a	*wari* are former Guajá and are eaten by Guajá; howler monkeys eat *yúa* palm fruit and are hunted by Yu-či'a
ka'i	brown capuchin	*aiyã*, Guajá, *wã'i* palm fruit	Yu-či'a, Mariawa-či'a	*ka'i* are made from *aiyã* or Guajá and are eaten by Guajá; *ka'i* are made from *wã'i* and eat *wã'i*; *ka'i* eat *yúa* and *mariawa* palm fruits and are hunted by Yu-či'a and Mariawa-či'a
ka'ihu	Ka'apor capuchin	*aiyã*, Guajá	Yu-či'a, Mariawa-či'a	*ka'ihu* are made from *aiyã* or Guajá and are eaten by Guajá; *ka'ihu* eat *yúa* and *mariawa* palm fruits and are hunted by Yu-či'a and Mariawa-či'a
kwič*ʰúa*	bearded saki	*awa*, *aiyã*, *kipi'i* termite mound, leaves	Mara'awi or Mariawi	*kwič*ʰ*úa* are made from *aiyã* or Guajá and are eaten by Guajá
yapayu	squirrel monkey	*mariawa* palm, *aiyã*, *awa*	Mariawa-či'a	*yapayu* are made from *aiyã* or Guajá and are eaten by Guajá; *yapayu* eat *mariawa* palm fruit and are hunted by Mariawa-či'a
aparikə	owl monkey	*aiyã*, Guajá, *akarahu* plant, leaves	Apariowa-či'a	*aparikə* are made from *aiyã* or Guajá and are eaten by Guajá; there is a possible etymological relation between the divinity Apariowa-či'a and the fruits *aparihu* and *aparikəwa'i* eaten by *aparikə*
etamari	golden-handed tamarin	*aiyã*, Guajá, *yahárə* leaves, *wã'i* leaves, *awynia*, *mariawa*	Yahárə	*etamari* are made from *aiyã* or Guajá and are hunted by Guajá; *etamari* are made from *yahárə* and are hunted by Yahárə; *etamari* reported to eat *mariawa*

TABLE 7.1 Cannibalistic Associations *(continued)*

Guajá Name	Common Name	Created From	Divinity	Endocannibalism
a'ɨ, a'ɨhu	sloths	*kipi'i* termite mound	Irahu	
akamu	porcupine	*kipi'i* termite mound	Irahu	
taminawa	anteater	*tapiki'i* termite mound	Takwariro	
tatu	armadillo	ears made from leaves	Manio, Karimičiči	
akuči	agouti	*aiyã*, *ka'a tata'ɨ* plant, *yahára* palm, *yúa* pod	Yamokware	*akuči* are made from *aiyã* and are eaten by Guajá; *akuči* are made from *yahára* and reported to eat *yahára*; *ka'atata'ɨ* may be Calathea sp. eaten by agouti
kararahu	paca	*aiyã*, *akarahu* plant, rock, *pinõwa* palm	Kamara-yu, Karai-yu	*kararahu* are made from *aiyã* and are eaten by Guajá; *kararahu* reported to eat *akarahu*; pacas eat palm fruit
tamakai	squirrel		Inaya, Yamokware	*tamakai* eat inaya and are hunted by Inaya
awynia	mouse		Yawãrə maraka'ia	
kapiawárə	capybara	Mai'ira's hair	Irapayu, Irapayárə	
tapeči	rabbit		Čape	*tapeči* reported to eat *čape* and are hunted by Čape
arapʰa	red brocket deer	*aiyã*, termite mound	Kamaraiyu, Karaiyu	*arapʰa* are made from *aiyã* and are eaten by Guajá
tiaho	white-lipped peccary	rock, termite mound	Panã-čia, Irapayárə	*tiaho* are reported to eat *panã* plant and are hunted by Panã-čia
mata	collared peccary	rock, *tamiki'i* termite mound	Aninaya or Aninayárə	
tapi'ira	tapir	rock	Kamara-yu, Karai-yu, Ipi-čia, Araku	
hairárə	tayra		Haira	

kwači	coatimundi	*ka'a pirahu* plant, *mukúrə*	Wipo'ĭ-čĭ'a or Ipo'ĭ	
yapuči	kinkajou	*aiyã*	Yurə, Apariawa-čĭ'a	*yapuči* are made from *aiyã* and are eaten by Guajá
mukúrə	opossum	*aiyã*	Yawárə-hu pinahun	
yawárə	jaguar	*yawa ka'a* plant, Mai'ira's head, wood	Čimírə, Kamai-yu	
yawatérə	otter			
kamičia	tortoise		Kiripi	*kamičia* are reported to eat *kiripi'i* palm fruit and are hunted by Kiripi
inama'i, maihu, tsikurihu	snakes	Makawɨ		*makawɨ* birds reported to eat snakes
yakare	caiman	wood, rock	Yawatérə	*yakare* reported to eat *yawatérə* and are hunted by Yawatérə
iramiri	small birds	small leaves	Yamokware, Kapitã	
inamõ	tinamous	big leaves	Yamokware	
mitõ	curassow		Uɨyárə	
takana	toucan	*yahárə*	Uɨyárə	*takána* are reported to eat *yahárə*
uru	vulture		Arawɨyárə	
wiraho	eagles and hawks		Uɨyárə	
wirahu	harpy eagle		Uɨyárə	
yaku	guans	*yahárə*	Uɨyárə	
arárə	parrots		Uɨyárə	
pira	fish		Yawačí-čĭ'a, Yawárə	*yawačí* are described as fish-eating birds
arawɨ	surubim catfish	*arama'a* plant		

All the divinities listed on table 7.1 specialize in hunting the associated animal, with the exception of Kiripi, who is a female divinity and is associated with the *kamičia* (the red-footed and yellow-footed tortoises). Some say that Kiripi is responsible for creating the *kamičia* and that she also collects and eats them. However, others say that *Kiripi* never eats *kamičia* but only collects them to keep them as pets. As previously mentioned, the female divinity Piraya keeps all species of monkeys as pets.

Partial consanguinity through the creation myth, spiritual siblingship, or *aiyā* predation creates relations of cannibalistic consumption among the Guajá, which can be summarized as the eating of a former version of the self, being eaten by a former version of the self, or reciprocal consumption by related forms of life. The "what eats you is what you were" theme is best expressed through the howlers. Howlers are former human beings who are eaten by human beings. The Guajá preferentially consume the game animal to which they consider themselves most closely related. For the other monkeys and many other game species, the same principle applies, but it involves two transformations—from human to *aiyā* to game animal. The "what you were is what you eat" theme is best expressed through the *aiyā* and plant origin myths for some game animals. The *aiyā* are former Guajá who eat the Guajá. For the brown capuchin, squirrel monkey, and several other game animals, the plant that they were created from is a plant that they consume.

Reciprocal predation ("what eats you is what you eat") is expressed through the *harəpihárə* spirit relationships the divinities have with plants and animals. Here, the animal is either made from the spirit *harəpihárə* of the divinity that hunts it, or the animal eats the spirit *harəpihárə* of the divinity that hunts it. Examples of the latter are: the divinity Yu-či'a, who eats howlers and other monkeys that in turn eat *yúa*; the divinity Panā-či'a, who eats white-lipped peccaries that in turn eat the *panā* (butterfly) plant; and the divinity Inaya who eats squirrels that in turn eat *inaya* palm fruit. Examples of the former are: the tamarin, who in one version is said to be made from *yahárə* and is also hunted by the divinity Yahárə; and the squirrel monkey, who is made from *mariawa* and is hunted by the divinity Mariawa-či'a. Further, the squirrel monkey also eats the *mariawa* palm, providing an example of both types of relationships.

Although a cannibalistic theme is apparent here in Guajá cosmogony and cosmology, it is not all-encompassing. For some animals, neither the creation myth nor the relationship of their controlling *karawa* divinity seems to relate to cannibalism. Some animals were made from the transformation of generic plant leaves and do not seem to have a special relationship with a particular divinity. Others have relationships with divinities, but symbolic cannibalism does not seem to be involved. For example, the divinity that hunts the coatimundi is called Wipo'ɨ̃-či'a, and is a spirit sibling with a vine. The coatimundi was not made from the vine, nor is it said to eat it. The same is true for the anteater, whose controlling divinity is Takwariro. The *takwariro* plant is a type of bamboo, and the Guajá do not believe that the anteater was made from it or eats it.

Often, when the spirit *harəpihárə* of a divinity is an animal rather than a plant, the divinity corresponds directly to the animal in a one-way predatory role rather than the more reciprocal cannibalistic exchanges characteristic of the plant relationships. For example, the sloths, *a'ɨ-te* and *aɨ-hu*, are hunted by the divinity Irahu, which is a term used for large birds such as the harpy eagles that prey on sloths (as well as monkeys and macaws). The Guajá describe a fish-eating bird called *yawači*, and the divinity Yawači-či'a controls fish. They also describe a snake-eating bird called *makawɨ*, and the divinity Makawɨ controls snakes. However, with the divinity Yawatérə, whose spirit *harəpihárə* is the otter, there is reciprocal cannibalism rather than one-way predation. Yawatérə controls the *yakare* (caiman), and the *yakare* will eat the otter.

Sources of Variation

Considerable variation existed in the interpretation of cosmological matters, which likely relates to the individualism characteristic of hunter-gathering peoples and to the nature of the *iwa* experience itself. First, as hunter-gatherers, family bands spend a good deal of time apart from each other, which would allow ample opportunity for drift in Guajá folklore. In addition, the Guajá have no headman or specialist in the group who could serve as an authority. Although the Guajá both respect the views of elders and rely on them to teach their stories, an elder's particular version of a story is not considered absolute. Guajá individualism also plays a role in this folkloric drift. When individual Guajá were asked about differences between their versions (such as the alternate stories of animal origins) and those of others, some individuals were firm that their own version was the correct one. Others allowed for all versions to be correct, stating, for example, that some animals were created by Mai'ira from multiple sources. The nature of the *iwa* experience itself likely generates much cosmological diversity. The *karawárə* ritual is highly individualistic; each man interprets his own personal visions to the group. The linking of the dream experience to the *iwa* for both men and women may create further variation. Although the Guajá may be able to consciously shape the visions seen in the *karawárə* ritual, they have less control over the content of their dreams, which might oblige the group to tolerate variant interpretations to some extent.

Bird Symbolism and Cannibalism

A connection may exist between the bird symbolism used by the Guajá in the *karawára* ritual and cannibalism. The dress for the *karawára* ceremony is very specific. Armbands and headbands are made from toucan feathers and the rest of the body is ornamented with vulture feather down. While the Guajá symbolically link themselves to birds when they say that they fly to the *iwa*, I was unable to evoke specific ideas about why the toucan and vulture feathers are used, apart from the assertion that "we do it this way because we are *awa*." (As previously mentioned, one of the reasons for excluding the *Aramaku* hunter-gatherers on the reserve from their group was that they wear parrot-feather headbands.)

Although the Guajá do not describe cannibalism in terms of bird symbolism, other evidence suggests at least the possibility that a cannibalistic link may have existed at some point in their culture history. In addition to the named *karawa* divinities associated with game animals, the Guajá describe many other *karawa* in the *iwa*. Generically, the food of the *karawa* is considered to be toucans. The specific dress of the Guajá for their journey to the *iwa* in toucan headbands and armbands suggested to me that they were becoming food for the divinities. Further, the adornment of their bodies with vulture feathers, a form of life that eats the dead, suggests a connecting between the *karawa* experience and the death experience. In addition, the vulture is one of the few foods that is universally taboo to the Guajá. In my view, it seemed that in dressing as toucans and vultures, the Guajá were sending the message that they are edible and inedible, simultaneously. The Guajá were going to the place of the dead without being dead. However, the Guajá denied that the *karawa* experience involved any notion of consumption by the divinities.

If either of my interpretations has any validity, it is possible that bird symbolism may have had a link to cannibalism at one time, and although it no longer does, the dress is continued due to its importance in tradition and group identity. But it may very well be that the practice has never had cannibalistic associations.

COMPARISON WITH SELECTED LOWLAND GROUPS

In addition to the proposition that animism is often a form of cannibalism and the examples given above, many other similarities exist between the symbolic cannibalism of the Guajá and cannibalistic practices and beliefs practiced among other groups, suggesting that cannibalism is a widespread organizing principle in Amazonian Indian cosmology. I am tempted to suggest that the relation between animism and cannibalism may be a logical analogue to the relation between totemism and taboo in terms of kinship relations.

The interrelations of ecology, kinship, and cosmology are key to understanding Guajá symbolic cannibalism. Among the Guajá, the role of monkeys, and particularly howlers, in their associations is likely more pronounced due to the monkeys' close physical and behavioral similarities to humans and their dietary importance. As previously described, other groups have linked monkey consumption to cannibalism, such as the Bororo (Crocker 1985), the Kalapalo (Basso 1973), and the Suyá (Seeger 1981) who value eating monkeys because of their similarity to humans while the Parintintin (Kracke 1978), Westernized Barí (Lizarralde 1997, 2002), and the Jo/'huasani (Schapera 1951) of southern Africa avoid eating monkeys for the same reason.

The *aiyā* and the *karawa* are familiar entities, often linked to symbolic cannibalism among various Tupí-Guaraní groups, although their interpretation varies.[2] Generally, other groups' versions of the *aiyā* are malevolent supernatural beings or forces, often associated with the dead, for example, *āñī* among the Araweté (Viveiros de Castro 1992), *ājā* among the Ka'apor (Balée 1984), *añang* among the Parintintin (Kracke 1978, 1999), *anhanga* among the Parakanã (Fausto 1997), and *anyang* or *ayā* among the Wayãpi (Campbell 1989, Grenand and Grenand 1981). Similar mythic creatures are also found in numerous non-Tupí-Guaraní groups, such as the *weorí mahā* of the Tukanoans and the mythic cannibal monsters in the Gê-speaking Suyá (Seeger 1981).

The Araweté's cosmology shares a number of similarities with Guajá's. Viveiros de Castro (1992) describes a parallel in their creation myth, where one of the twin sons seeks revenge on the jaguar for the death of his mother by transforming the guests at a maize-beer festival into the various forest animals. Here, the Araweté's game are former Araweté, suggesting a similar cannibalistic theme. Viveiros de Castro also describes that upon death, the souls of the Araweté are eaten by cannibal divinities, who then resurrect them from their bones. While the Guajá do not believe they are cannibalized by the divinities, as noted previously, Guajá bird symbolism suggests that they may have had this belief at one time.

The *āñī* of the Araweté, like the Guajá *aiyā*, are foul-smelling, earthbound cannibals who appear at night and are associated with the dead in that the "terrestrial specter of the deceased is said to accompany the *āñī*" (Viveiros de Castro 1992:68). Viveiros de Castro (1992: 288) further describes the *āñī* as "the bad memory of rotten flesh," which is said to generate the specter associated with the *āñī*. Thus, for the Araweté, the cannibal ghosts are specifically identified as past images and bad memories, which is possibly related to how the Guajá conceptualize the *aiyā* , given their ideas of past images of selves and others having a permanence in the *iwa*. One key difference between the *āñī* and the *aiyā* is that when the *āñī* cannibalize an individual, they cause complete death and prevent an individual from ever reaching the celestial home (Viveiros de Castro 1992). For the Guajá, the *aiyā* are instrumental in getting individuals to the *iwa*.

The Matsigenka of Peru have an entity that is similar to both the Guajá *hatikwáyta* and the *karawa*. Matsigenka shamans are believed to have a twin brother among the *Sangariite* (invisible ones), which are benevolent beings in the spirit world that raise the Matsigenka game as pets in their invisible villages and release them to be eaten by the people (Shepard 1997, 2002). This is similar to the Guajá idea of various *karawa* divinities having principal control over specific animals, which they will send near the Guajá if they are petitioned through ritual. In addition, when the shaman goes into a trance, he and his spiritual twin switch places so that the shaman visits the sky-village of his twin, petitioning the divinities to bring game, while the spiritual twin heals other members of the Matsigenka group (Shepard 1997, 2002). This is quite similar to the Guajá spirit possession experience during the *karawárə* ceremony. The idea of the twin is also similar to the Guajá *hatikwátya*, which are counterparts to the self in the spiritual world of the sky.

While the notion of the *karawa* in some groups is quite similar to that of the Guajá, in others this phenomenon is associated with the *aiyā*-like entities. Holmberg (1969) describes the *kurúkwa*, which may be cognate with the Guajá *karawa*, as ghosts of the dead that look like human beings and lurk in the woods waiting for victims to strangle. Similarly, among the Wayãpi, the term *kaluwa* expresses supernatural evil, and the term *kaluwaku* is synonymous with the notion of the *aiyā* ghosts of the dead (Grenand and Grenand 1981). The Parakanã have a notion of *karawa*, but it is related to their sentiments against cannibalism. According to Fausto (1997), the Parakanã avoid eating the paca, which is literally called *karowara*, because, mythologically, paca are believed to have been created from human beings, are *awa-kwera* (formerly human), and are associated with the nocturnal *anhanga*, cannibals similar to the Guajá *aiyā*. The Guajá similarly state that the howlers are *awa-kérə* (formerly human), and for that reason, they are the preferred food. The Parakanã terms are surely cognates with Guajá terms, but the meaning for the Guajá is inverted. This is also seen among the Ka'apor, where the *karuwar* are malignant spirits similar to the *ãjā* (Balée 1984:96–97, 174). The Ka'apor believe that the *ãjā*, along with another spiritual type of spiritual entity, the *ma'e*, cause all diseases.

The Kagwahív (Parintintin) notion of the *añang* (Kracke 1978) seems to have characteristics of both the Guajá *aiyā* and the *karawa*. Añang are similar to the *aiyā* in being ghosts that threaten human beings, in this case by making them crazy or leading them off into the jungle. But, according to Kracke, the *añang* are not earthbound, and they may visit the Kagwahív in dreams, which is similar to the Guajá idea of the *karawa* or *hatikwáyta*. Kagwahív shamans also travel to visit spirits in the sky, who are petitioned to heal the sick.

Similarities are also found among the Brazilian and French Guianan Wayãpi groups. The behavior of the *ayã* reported among the Wayãpi in French Guiana is quite similar to that of the Guajá (Grenand and Grenand 1982). Like the Guajá, the Wayãpi *ayã* are considered malevolent spirits who are the ultimate cause of all illness and death. They also behave identically to the Guajá *aiyã*; both groups believe that when a child falls down, it is due to these malevolent spirits pushing them or blowing them down. Among the Wayãpi, *anyang* is the term for grandfather as well as ghosts or a ghostly quality of the forest (Campbell 1989). Shamans have *anyang* which take the form of various animals and are said to be like pets inside of them. The Wayãpi also have a term, "*yarr*," which is likely cognate with the Guajá term *yárə*, a spiritual principle which is sometimes used synonymously with *hatikwáyta*. The Wayãpi *yarr* are personified spirits that inhabit the trees and are control various forms of life and may cause harm if angered. The term "*yarr*" is also used to describe the owner of a pet. The *yarr* seem to be similar to the Guajá *karawa* and *aiyã*. They are similar to the *karawa* in being controllers of various forms of life and, as with the Guajá, that role seems to be mirrored in pet-keeping, since the same term is used. However, the *yarr* are like the *aiyã* in being anthropomorphic forest dwellers and in having the potential to cause harm.

Mortuary cannibalistic practices and related beliefs among several groups resemble Guajá beliefs regarding the *aiyã*. According to Clastres (1974), the Aché (Guayaki) believe that a malevolent soul is created at death. Eating the dead prevents their souls from later entering the bodies of the living and killing them. Similarly, Dole (1972) describes mortuary cannibalism as a defense mechanism against the souls of the dead among the Amahuaca. Conklin (2001) has recently written an extensive account of Wari' mortuary cannibalism and offers a very different look at the function of cannibalism in that society.[3] For the Wari', "compassionate cannibalism" is a way of mourning the dead and coping with grief. Close affines are responsible for eating the deceased, though it is considered dangerous for consanguines to do so. Generally, affinity or meta-affinity seems to govern real or symbolic cannibal relations among lowland South American groups. At first glance, the Guajá may seem to represent a departure from this in that they prefer to consume those forms of life who share partial consanguinity. However, Conklin's description of closely interacting affines becoming consanguinealized (discussed in chapter four) may provide an analogy to preferential consumption of partial consanguines among the Guajá. In addition, the Warí are reincarnated as white-lipped peccaries who present themselves specifically to their consanguines to be eaten.

While the monkey can be said to serve as a reflexive symbol for the self among the Guajá, individual and group identity is complex. Guajá group identity is predicated on understanding its relationship to other equivalent natural and supernatural communities. The maternal same-sexed sibling term, *harəpiana*, is used for these natural groups, which suggests a similarity as well as an equality with the natural animal species and the supernatural *karawa* divinities. However, the Guajá identify themselves most closely with the howler community, whom they view as being true siblings, or *harəpihárə*, and with whom they share common human origins.

The Guajá as a community look at the howler monkey community and see themselves reflected, but the individual Guajá is, in truth, a plural concept by reason of its multiple *hatikwáyta*, the death *aiyã*, bilocation in the *iwa*, and the cohabitation of the body with the divinities in spirit possession. Through their *iwa* alters, Guajá individuals are simultaneously self and other, with past versions of themselves living out independent lives in the *iwa*, yet at the same time being very much part of the individual, so much so that women are able to see through the eyes of these alters. While the *hatikwáyta* concept allows, to some extent, for one's individual consciousness to perfuse multiple bodies, spirit possession by the divinities allows plural consciousnesses in a single body. Further, with death, an individual is permanently split asunder into two contrary life forms: the celestial-bound final version of the sacred *hatikwáyta* and the terrestrial-bound, time-bound, malevolent *aiyã* cannibals. Individual identity is an incredibly complex matter, as one's own self exists in multiple versions in space and time and in beings with diametrically opposed characteristics.

Guajá symbolic cannibalism is the process which links, integrates, and transforms the various forms of life in the Guajá cosmos. In describing the Tupinambá, Viveiros de Castro (1992:303) states, "Cannibalism is an animal critique of society and the desire for divinization," an assessment which also rings true for Guajá cannibalistic beliefs. Eating is not an act that merely satisfies hunger; it has the transformational power to make another sacred. Applying cannibalism to sociality might seem contradictory since *kin* suggests a cooperative relationship, and *cannibalism* suggests a predatory relationship. However, in the broader scheme, cannibalism can perhaps be thought of as delayed reciprocity if all the plants, animals, and supernatural forms can be linked into a overarching social group. Rather than simple predator-prey relationships, consumption gives the consumed the gift of immortality, and the consumer knows that the gift will, at a later time, be reciprocated by another.

Conclusion

ETHNOPRIMATOLOGY IN AMAZONIA AND BEYOND

Although there are no living nonhuman-primate species endemic to the United States or Europe (except the Barbary macaque), they have had an important role in Western society through both science and popular culture. The focus of primatology in anthropology seems to be slowly shifting, or at least expanding from its original evolutionary orientation to a focus on purer ethology and primate conservation. In part, this is due to very real concerns about the fate of monkeys and apes whose continued existence is threatened by human behaviors.

ETHNOPRIMATOLOGY IN WESTERN PERSPECTIVE

Taking a step back, the earliest perspective on studying nonhuman primates was the notion that they could serve, in a sense, as living fossils. In many ways, this perspective is not all that different from the cultural evolutionary paradigm of the late nineteenth century, when all human groups were classified as falling into one of the three stages of savagery, barbarism, or civilization. In hindsight, it is clear that these views were at best ethnocentric and at worst reflected the racism, colonialism, sexism, and virtually ever other -ism of the times. However, it is important to remember that these early anthropologists were thinking out of their own cultural context. They presented the radical view that we should consider other peoples to be just as human as members of Western society rather than inferior nodes on the Great Chain of Being or degenerated subhumans who had fallen from the grace of God.

From an evolutionary perspective, the nonhuman primates have served as a kind of ethnographic retrodiction in the same way that "savages" did in the cultural-evolutionist paradigm. Although nonhuman primates are as much the product of their evolutionary history as are human primates, there has been a pervasive ideology that nonhuman primates are living fossils that represent an arrested stage of human development. The popular "ascent of man" image suggests that the knuckle-walking chimpanzee represents a prehuman stage in our development from scoliotic to erect spine, from small brain to large brain, and from nature to culture. While science does not suggest that humans evolved from chimpanzees, the notion seems entrenched in the popular imagination that humans are the final product of evolutionary forces, rather than a coproduct and that chimpanzees are a time capsule for humanity.

Today we are witnessing a redefinition of the relationship between human and nonhuman primates, particularly in the case of the chimpanzee. A hundred years from now, the isms will be clear, but at this point, it is not entirely clear what the eventual outcome will be. The question of the relationship of nonhuman primates to humans seems to pulling in two different directions at once. On the one hand, biological and behavioral evidence suggest the distance between humans and chimpanzees is much narrower than previously suspected. One consequence has been an invigoration of the animal rights movement, with apologists making arguments similar to those used in arguing for human civil rights. On the other hand, some seek to stress the distance, often in terms of religious doctrine, to cement human eminence.

My own home state of Alabama demonstrates some aspects of this pull in different directions. Researchers at the university where I work, the University of Alabama at Birmingham, were the first to find the link between human HIV and chimpanzees (Gao et al. 1999). On our campus, we have recently built a 37-million-dollar Human Genetics research facility with one research goal being the study of the evolutionary history of retroviruses. At the same time, since 1995 stickers have been pasted in Alabama K–12 textbooks which proclaim:

> This textbook discusses evolution, a controversial theory some scientists present as a scientific explanation for the origin of living things, such as plants, animals and humans. No one was present when life first appeared on Earth. Therefore, any statement about life's origins should be considered a theory.

In 2001, the Alabama Board of Education voted to include an expanded version of this disclaimer in the preface to state textbooks. Describing evolution as a theory, rather than fact, and one espoused by "some" scientists, suggests that these scientists represent a radical atheistic fringe. In our 1998 gubernatorial campaign, human evolution became an issue between the two key republican contenders. Winton Blount called the incumbent governor, Fob James, an embarrassment to the state because he had "danced like a monkey" in front of the school board to mock evolutionary science. James's reply was, in essence, "I may dance like a monkey, but you are a fat monkey."

For my own part, I must admit that there is a part of me that dreads teaching evolution in my Anthropology 101 courses. Inevitably, there is one in the crowd who has obtained his or her science from cable television—"Carbon-14 dating has been disproved." "How do you explain Job's description of dinosaurs?" "If evolution is true, why is there six inches of dust on the moon?" (I still haven't figured that one out.) But perhaps more disturbing than the challenges on pseudoscientific grounds are the concerns of those who fear for my mortal soul and stop by during office hours to witness and leave pamphlets or the sympathetic response of one student: "Don't worry, I know that you don't believe this and they are making you teach it." I am realizing that the true test of cultural relativism is not in understanding the belief systems of exotic others, but in having respect for the cultural beliefs of members of one's own society. And yet, it moves. A degree of distortion exists on both sides. Humanity is no more demigod than it is demichimpanzee.

However, over the last fifty years, it is not merely on ideological grounds that people question the distinctiveness of the criteria separating human beings from primates; studies in ethology and molecular biology have also left their mark. These studies have blurred the boundary, particularly between human beings and the great apes. Molecular evidence demonstrates a close link between human beings, chimpanzees, and gorillas. Sibley and Ahlquist's (1984, 1987) much-cited work in DNA hybridization found that human beings and chimpanzees had less than 2 percent genetic difference and that they were more closely related to each other than either was to the gorilla. Similar results were found by Caccone and Powell (1989). However, it should be noted that Sibley and Ahlquist's methodology has been criticized (Marks, Schmid, and Sarich 1988; Sarich, Schmid, and Marks 1989) and that previous and subsequent research has suggested which of the three species are closest is difficult to determine (e.g., Hoyer et al. 1972; Marks 1992; O'Brien et al. 1985; Sibley, Comstock, and Ahlquist 1990), and a rather heated debate ensued (e.g., Marks 1993; Sibley and Ahlquist 1993). But, setting aside the debate, what is clear is that the molecular data demonstrate an extremely close link between human beings and gorillas and chimpanzees.

Behavioral research has also blurred the line. Research on chimpanzee traditions are broadening the definition of what culture itself means. Whiten et al. (1999) described thirty-nine differing behavioral traditions among chimpanzee populations in areas such as tool-making, grooming, and mating behaviors. The study is a collaboration of researchers from the seven best-studied chimpanzee sites in Africa, and the combined research totals over 150 years of work. The differing behavioral traditions cannot be explained by genetic (subspecies) differences or variations in the ecological setting. Thus it suggests that chimpanzees have culture. In addition, social learning seems to play a role in zoopharmaconosy, the use of medicinal plants, among the African great apes (see Huffman 1997). It should also be mentioned that the New World capuchins demonstrate a range of tool-use behaviors that arguably rival that of the chimpanzees (e.g. Visalberghi and McGrew 1997; also see the appendix to this volume).

There has also been a great fascination with the ability of the great apes to learn American Sign Language to a limited degree. Chimpanzees have used it to converse with humans and to converse with each other, and can pass it on to the next generation in experimental settings (see Fouts, Fouts, and Waters 2002). This "limited degree" should neither be underemphasized nor overemphasized. The great apes do not seem to possess the hard wiring to use complex grammatical structures; however, enough cognitive similarity exists that humans and apes can, in a sense, speak to one another.

The molecular evidence, cultural traditions, and linguistic abilities among the great apes have lead some to argue for shifting our position in the classification of the great apes. Diamond (1993:21), for example, argues that the bonobo and the common chimpanzee should be included in the genus *Homo* and that human beings are best classified as a third species of chimpanzee. Cavalieri, Singer, and the contributors to the volume *The Great Ape Project: Equality Beyond Humanity* (1993) presented "A Declaration on Great Apes," arguing for the extension of human rights to the great apes. It begins, "We demand the extension of the community of equals to include all great apes: human beings, chimpanzees, gorillas, and orang-utans" (1993:4), which would include the right to life, the protection of individual liberty, and the prohibition of torture. Several have equated the treatment and perception of the great apes to the European abuse and enslavement of Africans (Nishida 1993; Teleki 1993; Cavalieri and Singer 1993). Groups such as GRASP (Great Ape Standing and Personhood) and GAP (Great Ape Project International) hold similar views on extending human rights to the great apes.

Criticism has ensued. Arguably, we are still dealing with ensuring basic human rights to all human beings around the globe. Criticism has also been directed at the messenger, especially Singer's (1993, 2001) controversial views on euthanasia and bestiality. One of the original signers of the Declaration on Great Apes, Rutgers professor Gary Francione, called for Singer's resignation as president of GAP, due to concerns that his position on bestiality could be used to justify the sexual abuse of great apes.

Perhaps the most controversial area involves the role of primates in medical research. Due to the close genetic relationship between humans, chimpanzees, and primates more generally, nonhuman primates have often been the species of choice for biomedical research. One recent example is the development of a vaccine against the Ebola virus for the cynomolgus monkey (*Macaca fascicularis*) that is hoped to have potential for developing into a vaccine against Ebola in human beings (Sullivan et al. 2000).

However, Fouts, Fouts, and Waters (2002) have described a number of problems with generalizing from nonhuman primates to human beings in biomedical research. One problem is that the life history of the primate is often not taken into consideration. The psychological stress of raising a social primate in isolation in a cage can result in neurological, endocrinological, and immunological physical abnormalities. Fouts, Fouts, and Waters have expressed particular concern about the use of chimpanzees in AIDS research. First, with the exception of one case,[1] chimpanzees who are infected with HIV do not respond with the full suite of symptoms that characterize human AIDS. One explanation they give is that chimpanzees have apparently co-evolved with SIV over at least the last several thousand years and have natural defenses for fighting off this type of virus. Fouts, Fouts, and Waters also point out that this makes it difficult to test the efficacy of a vaccine, because it is difficult to disambiguate the effects of the vaccine from the effects of the chimpanzee natural immune response. One additional problem they describe is that the close genetic relationship between chimpanzees and human beings may give the impression that SIV and HIV are closely related. However, due to the high mutation rates of these viruses, HIV-1 (the most common strain infecting humans) is only 40 percent homologous with SIV (Fouts, Fouts, and Waters 2002:54–55, citing Greek and Greek 2000).

The major objection to the use of nonhuman primates in research does not rest with their appropriateness as models for human disease, but rather with the ethics of research on monkeys and apes. A number of animal rights activist groups have been putting increasing pressure on research labs to eliminate the use of all nonhuman primates. The only research institute in Europe (the Biomedical Primate Research Center in Rijswijk, the Netherlands) conducting medical experiments (AIDS, hepatitis C) on chimpanzees is phasing out all research due to criticism from animal rights groups (Goodman 2001). Shamrock Farm, Britain's only supplier of nonhuman primates for research (mainly macaques) was recently forced to shut down after a campaign by animal rights activists (Aldous 2000). In 1997, the Coulston-operated Laboratory for Experimental Medicine and Surgery in Primates in New York was forced to close, due in part to protests from animal rights activists (MacIlwain 1997). At the Coulston Foundation in New Mexico, the NIH confiscated 288 chimpanzees (about half the population) after protests from animal rights groups regarding the poor living conditions at the facility (Smaglik 2000).

The darker side of this issue is seen when animal rights extremists resort to terrorist tactics to achieve their aims. The core should not be judged by the periphery, and the vast majority of animal rights activists use peaceful means of protests. However, a few do not. The Animal Liberation Front openly endorses vandalism and arson to inflict economic damage. In 1999, they claimed responsibility for fire-bombing four vans belonging to a Rhode Island furrier (Nadis 1999). In their "Year-End Direct Action Report" for 2001, they claim responsibility for seventy-two illegal actions in North America, including over a million dollars in arson damage at the Coulston Foundation (NALF 2002). Two radical offshoots of the Animal Liberation Front, the Animal Rights Militia and the "Justice Department" have been linked to more direct violence. Oxford physiologist Colin Blakemore claims to have been beaten by animal rights activists, and the "Justice Department" claimed responsibility for sending dozens of letters booby-trapped with razors to primatologists (Koenig 1999; Nadis 1999).

Another controversy involving medical research involves xenotransplantation, cross-species organ transfer. One of the earliest occurred in 1963 when chimpanzee kidneys were transplanted into six patients with chronic kidney failure (Nowak 1994). Perhaps the most publicized case occurred in 1984, when a baboon heart was transplanted into a fifteen-month-old human infant, "Baby Fae," who survived for thirty-five days (Nowak 1994).[2] Experiments have also been done with liver (for hepatitis) and bone marrow transplants (for AIDS) from baboons to humans (Bailey and Gundry 1997; Murphy 1996; Starlz et al. 1993).

Today, more than 50,000 people await organ donations in North America, and nonhuman primates have been considered a possible source (Bailey and Gundry 1997). But the need for organs far outstrips the supply. For example, the world's largest captive baboon colony, at the Southwest Foundation for Biomedical Research in San Antonio, has only 2,700 baboons, but there are as many as 43,000 people in the U.S. with heart failure who are unable to receive a transplant because of availability (Nowak 1994). To meet the demand, a mass breeding program would be needed just for heart transplants.

Recently, the controversy over xenotransplants has not been so much with ethics, but with the fear of disease transmission. The larger concern is that a zoonotic infection in an individual receiving a transplant could become infectious to other human beings, developing into another HIV-like epidemic in the human population (Murphy 1996). In 1999, the U.S. Food and Drug Administration suggested a moratorium on organ transplants from nonhumans to humans and in May of 2000, a Secretary's Advisory Committee on Xenotransplantation (SACX) was created under the U.S. Department of Health and Human Services to explore this problem (Butler 1999, 2000).

The present state of the nonhuman primates can be described in no terms other than a crisis of potential extinction for multiple species. There is no doubt that one of the primary causes is the devastation of forests by human beings. The term "biocultural diversity" has recently gained popularity in linking the fate of the forest to the fate of the endemic species to the fate of the peoples of the forest. What the term suggests is that biodiversity and cultural diversity are intimately linked. Essentially, the commodification of nature into resources to be bought and sold does more than transform the landscape, it transforms cultures. It becomes more than a question of the pros and cons of resource extraction. It is a question of conscience, which is complicated.

A problem exists when we project our own troubled conscience onto indigenous peoples. As Descola (1998) has so thoroughly described, in Amazonia (and elsewhere), there has been a tendency to use indigenous societies as a medium to project Western values related to animal rights and conservation, which very often bear little or no relation to the actual beliefs of the peoples themselves. In the case of Amazonian animism, plants and animals are often viewed as "persons." However, in Amazonian thought, personhood in no way suggests that animals are not legitimate prey. Arguably, Amazonian peoples have had little need to be conservationists except in the context of the encroachment of Western society.

Rose (2002) is critical of primate-conservation approaches that focus on protecting forests from people as well as approaches that stress sustainable development as a strategy. Rose argues that conservation needs to address not only local practices, but the regional and global economic and political conditions that impinge upon groups. For example, the civil war in the Republic of Congo is a serious threat to the bonobos, *Pan paniscus*. The population was estimated at 10,000 in 1996 but is believed to have been halved in the last several years as a result of outright habitat loss, habitat disturbance due to armed conflict, and increased hunting for meat (Saegusa 2000).

With regard to the latter, the major threat to primate populations is usually understood as a result of habitat destruction. However, in Africa, commercial primate hunting is now posing the more serious threat. As Rose sardonically states, "the explosion of the commercial bushmeat hunting is destroying primate populations faster that their habitats can be cut down" (2002:208). He argues that the situation is so critical in equatorial Africa that the conditions for sustainable indigenous primate hunting no longer exist. Further, he questions the validity of the term "traditional" as applied to hunting in groups that are increasingly replacing indigenous hunting technologies with guns and who are lured into the cash economy of the commercial bushmeat harvesting. He is even doubtful that traditional, sustainable primate hunting exists anywhere in the world.

Wieczkowski and Mbora (2002) provide an important cautionary tale for the premise that creating primate reserves protected from human influence solves the problem. According to them, in 1996 the World Bank funded a voluntary relocation program for the local people to move them from the area of the TRPNR (Tana River Primate National Reserve) in Kenya. The reserve is home to eight primate species including the critically endangered *Procolobus ruformitratus*. The problem was not only that the local people rely on the forest products for food, construction materials, and household items, but that they did not want to leave their homes. Wiezkowski and Mbora found evidence that the local people were willfully destroying the forest with the logic that if there was no forest, there would be no monkeys, and then, no pressure to leave their homes.

Fuentes (2002) also provides an example of the complexity of the conflicts among primate conservation, local culture, and the global economy. The Mentawai Islanders traditionally utilized the forest for numerous resources, including primate hunting. Commercial logging and the islanders' shift from subsistence to cash crops has dramatically reduced the primate habitat, with primates increasingly raiding gardens as a result. The reduction in the primate population has led to a ban by the Indonesian government on primate hunting in an attempt to protect the primates. While the ban on primate hunting helps to some degree, it misses the root of the problem. However, it is possible to develop programs that involve local people that do work. One success story is a Vietnamese project that involves local people acting as educators and rangers to protect the critically endangered Tonkin snub-nosed monkey, *Rhinopithecus avunculus*) (Eudey 2002).

One additional issue is that while primate conservation is of paramount importance, nonhuman primates have sometimes served as poster children for environmental concerns. The habitats that contain primates have often received attention that other threatened habitats have not. Saying this is not meant to diminish in any way the importance of primate conservation. Rather, it is meant to point out that the problem of primate conservation is part of a bigger picture of global change and that the value that Westerners place on nonhuman primates can lead to greater attention to their specific habitats. My own home state of Alabama (see Cormier 2002b) ranks near the top in both biodiversity and the number of threatened species in the United States. However, the species which are most threatened are river mollusks and snails, and the Save the Snails campaign has not been effective in garnering widespread public support for environmental conservation. But the loss of the river is a symptom of changes in river ecology due to pollution, which is itself a symptom of urban sprawl. While I believe that endangered primate habitats should receive top priority, it is important not to forget that the problem is part of a bigger picture of global environmental changes affecting numerous endemic species and cultures.

HISTORICAL ECOLOGY

The historical ecological perspective takes into account the mutual influence of culture change and environmental change over time. A functionalist or ecologically deterministic approach is insufficient for understanding the Guajá. Even attempting to attach a label to them as traditional hunter-gatherers is not wholly accurate, for it only partially captures them in a particular moment of time. The evidence suggests that the Guajá of the past were a horticultural people uprooted from their habitat and pushed eastward by one or more of the multiple effects of enslavement, disease, and warfare. Thus, their foraging "tradition" with its intense focus on monkey hunting is unlikely to extend back more than several hundred years. If the ongoing deforestation of their habitat continues, it may not extend much longer into the future.

My own research was possible because of the way the Guajá are moving away from their foraging lifestyle. The Carajás railway, among other factors, has led to the destruction and circumscription of their foraging grounds, which consequently led to FUNAI's creation of their indigenous reserves and their increased contact with non-Guajá people. Today, some are raising crops rather than being full-time foragers, some are shooting monkeys with guns rather than bows and arrows, and they exposed to or adopting, or both, many Western cultural items. As described in the introduction, they are foragers who sometimes watch American television via FUNAI's satellite dish. More importantly, the changes to their habitat are ongoing. While I believe that the primate hunting they currently engage in is sustainable, I am not optimistic that it will continue to be so if the current trends persist.

I have described monkeys as playing a multidimensional role in the Guajá culture. Their ecological adaptation is unlikely to reflect an ancient history as hunter-gatherers; rather, it is a consequence of historical events leading to their loss of agriculture and adaptation to ecological environments created in anthropogenic forests. Monkeys are the key game source in the wet season and trekking behavior is related to their use of monkeys as food. Prior to sustained contact, the Guajá were known to move from old fallow forest patch to old fallow forest patch, where they relied heavily on the babassu palm for food. Such a location attracts a number of monkey species that also exploit old fallow plants. In addition, the importance of monkeys in the diet is reflected in Guajá ethnobotanical knowledge. The Guajá know more about plants that are edible for game, and particularly the monkeys, than they know about plants edible for humans. Knowledge of game plants is, then, an important hunting strategy.

Guajá social organization and cosmology bear much in common with those of other Amazonian peoples, particularly those of the Tupí-Guaraní speakers. Guajá animistic and symbolic cannibalistic beliefs regarding their relationship to the natural and supernatural world reflect similar constructions in other groups. However their specific cultural configuration, involving extending siblingship terms to nonhuman others, differs from that of many groups, and seems to reflect the basically egalitarian nature of their society, characteristic of foragers. The focus on primates, and particularly howlers, is also unique in the region. Howlers are believed to be former humans and are believed to be the nonhuman community most closely related to the Guajá. The apparent contradiction between monkey as kin and monkey as food is logical in terms of Guajá ecology, social organization, and cosmological beliefs. Most forms of life are part predator and part prey, interconnected in the same kinship matrix. Consumption is a biological, social, and cosmological act whereby those consumed are transformed and transported to the sacred *iwa*.

While Guajá beliefs resemble that of other Amazonians, they are constituted in their particular means of adapting to the environment. However, ecological determinism is insufficient to explain Guajá interactions with monkeys. Although the availability of monkeys is obviously a prerequisite for their exploitation, other cultures in the eastern Amazonian region have not adopted a foraging mode of production nor do they exploit monkeys to the extent of the Guajá. It is doubtful that the Guajá originally adopted a foraging lifestyle voluntarily. The loss of fire-making ability suggests that a catastrophe befell these people, leading to cultural loss in a number of areas. Although the historical record of early contact with the Guajá is meager, the tragedies of their recent history likely have parallels in the colonial past: killing by neo-Brazilians, survivors scattered hundreds of miles apart, and devastation by introduced diseases.

Despite the historical events that altered the way of life of the Guajá, they have survived and adapted. Their foraging mode of production is as much an adaptation to the environment as it is to social conditions, not only because of the damage done by neo-Brazilians, but because of Guajá exploitation of old fallow forests, which suggests that the presence of other horticultural cultures is prerequisite to their survival as a foraging people. In this light, it is tempting to characterize the Guajá as deculturated. However, despite their losses, their foraging way of life suggests to me "reculturation." Although historical events likely led them to this way of life, monkeys have become integrated into their foraging behavior, their social system, and their cosmological beliefs.

When considering the massive habitat destruction occurring throughout the Amazon, the relationship that the Guajá have with local monkeys can be considered to be mutually beneficial. In their area, the real danger to the monkeys lies in the deforestation of the indigenous areas of the Guajá. The habitat of the newly identified *Cebus kaapori* has been so delimited that it only exists in the vicinity of the Guajá reserves, with perhaps a patchy existence to the west. The habitat of the endangered subspecies *Chiropotes satanas satanas* is almost as restricted, with the reserves providing a refuge from deforestation.

Despite the mutual benefits of the relationship between the monkeys and the Guajá people, the Guajá cannot be considered ecologically noble savages. In recent years, their way of life has begun to change rapidly. Although the *iwa* represents cultural purity, and some Guajá still shun contact with outsiders, their exposure to non-Indians has created new desires in them. Western goods, such as steel tools, guns, and ammunition, are becoming increasingly important to them. They have even asked FUNAI for a car since learning the Guajajara have one, and they have been upset that it has not been delivered, despite the complete uselessness of a car in the middle of the forest. Some complain about extended treks because there is no salt, no manioc, and no Western medicine deep in the forest. It would be inappropriate to blame contact with the FUNAI for their acculturation. I am convinced that without the intervention of the FUNAI, the Guajá would not exist today as an independent culture. Further, the officials at the P.I. Awá are extremely courageous to stay among the Guajá people, despite the repeated death threats against them by locals who want the Guajá land.

The biggest threats to the Guajá culture are disease and invasion of their habitat. Available FUNAI medical records from 1989 to 1990 indicate that approximately 30 percent of the population suffers from malaria each year, with approximately 23 percent of the cases being falciparum malaria, which was recently introduced to the area (Coelho 1994a, 1994b, 1994c; Damasceno da Silva 1996a, 1996b, 1996c, 1996d; Tomaz Filho 1991a, 1991b, 1991c; Nobre de Madeira

1988, 1990; Santos 1990). Viral respiratory infections also plague the Indians. These medical reports also indicate repeated infections affecting the entire group, particularly in the wet season. I observed the same pattern during my stay with the Guajá. Two of the Guajá have been diagnosed with tuberculosis, and one of them died during the research period.

Destruction of the Guajá habitat has continued unabated. FUNAI reports repeatedly lament insufficient personnel to deal with the constant invasions by non-Indians (e.g., Damasceno da Silva 1995a, 1995b, 1995c; Tomaz Filho 1991b, 1991c). The train has facilitated the growth of nearby villages and the infiltration of the reserves by *posseiros* (illegal squatters) as well as loggers. Two of the chief commercial trees logged are *Manilkara huberi* and *Tabebuia* sp., both of which are culturally important to the Guajá. *Manilkara huberi* is reportedly heavily utilized by the howlers, and in the Guajá creation myth, it is the species human beings were eating before they were transformed into howlers. *Tabebuia* sp. is the only tree used by the Guajá to make their bows.

FUNAI officials have repeatedly attempted to reason with locals not to hunt, fish, or settle on the reserves, not only because it is illegal, but because the Guajá will kill invaders. The response of the locals is that they have to feed their hungry children, and they feel that they have no choice but to take the risk. The Brazilian locals face extreme poverty, and although my contact with them was limited, their lives are likely similar to the desperation among other Northeastern Brazilian villagers, as described by Scheper-Hughes in *Death Without Weeping* (1992). It is difficult to find a villain among the Guajá and the local Brazilians, who are two peoples both trying to survive in an area where land is at a premium.

The situation is complicated, but it is clear that Brazil's economic problems and the poverty of many of the people will not be solved by converting the Guajá land into rice paddies and cattle pasture. Preservation of the reserves is essential for the survival of the Guajá culture, as well as for the literal survival of many of the monkeys and other endemic plant and animal species. The relationship of my European ancestors with Native Americans in the United States leaves me little room to criticize Brazilian policies. In fact, the involvement of the World Bank in the CVRD project makes me personally a part of the process every time I pay taxes. What I do hope is that the research herein can contribute in a small way to recognizing the value of preserving the forest, the plant and animal species within, and the unique culture of the Guajá people.

Appendix

MONKEYS IN THE GUAJÁ HABITAT

Seven species of primates are found in the Guajá area. The following provides a brief review of some of the studies by primatologists on these species. While the review of the literature provided here is not exhaustive, it provides an overview of some of the studies on the behavior and ecology the monkeys in the Guajá habitat. It should also be mentioned that there is an inherent unevenness in the descriptions, for some species are far better known than others.

ALOUATTA BELZEBUL BELZEBUL
(BLACK AND RED HOWLER, RED-HANDED HOWLING MONKEY)

Taxonomy and Distribution

According to Mittermeier and Coimbra-Filho (1981), some debate has existed regarding the classification of *Alouatta* at both the species and subspecies level. Many subspecies are poorly defined, and their relationships are not clear. Some debate exists as to whether A. *palliata pigra* is best considered a subspecies of A. *palliata* or a separate species. The name A. *villosa* has created some confusion since it has also been applied to A. *pigra*. Confusion also exists with the name A.

guariba, which is used by some for the brown howler. Mittermeier and Coimbra-Filho suggest distinguishing six howler species: A. *belzebul*, A. *caraya*, A. *fusca*, A. *palliata*, A. *seniculus*, and A. *villosa* (*pigra*). A. *belzebul* has a wide, but disjunct distribution in Northeastern Brazil (Almeida, Pimentel, and Silva 1995; Coimbra-Filho, Câmara, and Rylands 1995). Bonvicino, Langguth, and Mittermeier (1989) have divided A. *belzebul* into four subspecies based on their geographic distribution and the dominance of pelage pattern. All A. *belzebul* subspecies are located in eastern Amazonia and the northeastern Atlantic forest with A. *belzebul belzebul* found in several areas in northeast Brazil, including the northeastern portion of the Atlantic Forest region of eastern Brazil from the Paraíba to Alagoas, south of the latitude of Marabá (5°21'S), east of Serra dos Carajás, and west of the 45th parallel and the basin of the middle Rio Tocantins.

Physical Appearance

Crockett and Eisenberg (1986) have described the physical characteristics of *Alouatta*. They are among the largest of the Neotropical primates with an adult size ranging from 4–10 kg. Howlers are distinctive in having an enlarged hyoid bone which acts as a resonating chamber and amplifier when the monkeys vocalize to advertise territory. Howlers are sexually dimorphic in both body size and in their hyoid bone size. They possess a prehensile tail, which is used for support in feeding and for bridging gaps in the vegetation.

The pelage pattern of the subspecies A. *b. belzebul* has been described by Bonvicino, Langguth, and Mittermeier (1989). The pelage is both rufous and black. Two types exist. In the dominant type (71.8 percent), the coat is black, but the hands, feet, and end of the tail are rufous. The second type (28.3 percent) looks like the first except that the top of the head or nape is also rufous. In the Guajá area, both types were seen with the dominant type more often seen than the second. In addition, Schneider et al. (1991) have observed completely red variants of A. *belzebul* east of the Tocantins. As noted in chapter five, one informant reported seeing a completely reddish howler in their area.

Habitat, Locomotion, and Diet

The diet of A. *belzebul belzebul* in the Atlantic forest varies seasonally. In the wet season, 95 percent of the time was spent eating fruit, and in the dry season,

43.9 percent of the time was spent eating fruit, with an average of 59.0 percent of the time spent eating fruit for both seasons (Bonvicino 1989). An important tree in the diet of *Alouatta belzebul* is of the *Virola* genus, which is also the tree genus most widely used by loggers as float trees to raft logs of heavier trees downstream (Marsh, Johns, and Ayres 1987). *Manilkara huberi*, another key food source, is also selectively logged in Maranhão (Johns and Ayres 1987).

Little information is available on *Alouatta belzebul*, but other howler monkey research may be instructive. In general, howlers are able to exploit a wide range of forest types, including lowland evergreen forest, highland forest, swamp forest, and riparian forests (Milton 1984), as well as secondary forest and disturbed habitats, as previously described. Fleagle and Mittermeier (1980) and Mittermeier and van Roosmalen (1981) have described *A. seniculus* in Surinam. *A. seniculus* was most frequently found in the middle and upper stories of the main canopy and in the emergent trees of the high forest, but it was found occasionally in the low rain forest, mountain savanna forest, liane forest, pina swamp forest, and edge habitats. *A. seniculus* used predominantly quadrupedal locomotion during travel (80 percent). Climbing (16 percent) and some leaping (4 percent) were also used. During feeding, only about 56 percent of locomotion was quadrupedal. Howlers make extensive use of their prehensile tail. The howler was predominantly folivorous, eating a variety of leaves, including mature leaves, flowers, buds, and fruit. They did not observe insect-eating or seed-eating.

Activity Patterns and Range

Bonvicino (1989) found that the activity budget of *A. b. belzebul* in the Atlantic forest of Brazil was as follows: time spent resting (54.3–60.5 percent), travelling (19.9–22.4 percent), eating (2.5–11.9 percent), vocalizing (3.4–34.6 percent), grooming (0.4–0.7 percent), and playing (2.9–5.2 percent). Home range data was documented for two groups. One group had a home range of 9.5 hectares and the other had a home range of 4.75 hectares. Howlers in general spend a substantial amount of time at rest. Smith (1977) found that *Alouatta palliata* on Barro Colorado Island spent 21 percent of the time moving or feeding, 79 percent of the time motionless or autogrooming, and less than 1 percent of the time involved in overt social interaction. *Alouatta villosa pigra* in the National Park of Tikal spent 66 percent of the time resting, 12 percent of the time moving, and 22 percent of the time feeding and foraging (Schlichte 1977).

Social Organization and Social Behavior

A. *belzebul belzebul* in the Atlantic forest of Brazil demonstrated a group size ranging from five to fourteen individuals for five groups (Bonvicino 1989). However, this habitat is extremely threatened. Clarke and Zucker's (1994) work with *Alouatta palliata* in Costa Rica, as previously mentioned, suggested that environmental change can alter group size and behavior. Crockett and Eisenberg (1986) documented a typical pattern of female howler emigration; however, Clarke and Glander (1984) found that A. *palliata* groups consisted mostly of unrelated males and females, and both male and female juveniles emigrated.

A. *palliata* on Barro Colorado Island formed multi-male troops but had a minimal amount of overt social behavior (Smith 1977). Clarke and Glander (1984) described almost linear hierarchies among A. *palliata* based on access to food and resting sites. Crockett and Eisenberg (1986) stated that, in general, howlers spend less than 2 percent of the time grooming. Intergroup communication is more obvious. Daytime howls are used to advertise territory to other troops (Crockett and Eisenberg 1986). Smith (1977) suggested that howling and low levels of overt social activity are adaptations to a folivorous diet. He argued that howling is used to avoid potential conflict with other troops. The low level of overt social interaction within the troop among *Alouatta palliata* was argued to be adaptive in that it allowed individuals to minimize the use of their voluntary musculature. Thus, their bodies were able to concentrate physiological activity on their diet of readily obtainable, but difficult to digest foods.

Several stereotyped behavior patterns have been reported for several howler species (Crockett and Eisenberg 1986). These include tongue flicking, chin-throat-chest rubbing, back rubbing, anogenital rubbing, and the arch display. The arch display is an aggressive posture in which the back is arched and head lowered, sometimes accompanied by piloerection and branch breaking.

Reproduction

Howlers are not seasonal breeders, and births can occur in any month (Crockett and Eisenberg 1986). The average age of female sexual maturation is 187 months (Happel, Noss, and Marsh 1987). However, Crockett and Eisenberg note that the age of sexual maturation can be affected by environmental nutrition (1986). They also give further data on howler reproduction. Interbirth intervals after surviving offspring average 11–24 months. The range for gestation is 180–194 days and estrous cycles average 16–20 days.

Clarke and Glander (1984) have identified factors involved in female reproductive success among *Alouatta palliata* in Costa Rica. Clear sex differences
were seen for infant survival. Among infants who died naturally or survived beyond one year of age, 100 percent of the females survived compared to 39 percent of the males. The presence of a peer within three months of age affected
male survival. Survival rate was 56 percent for those with a peer and 20 percent
for those without a peer. Survival rate was also significantly higher if a sibling
was present in the group. Male takeovers also adversely affected male infant
survival. Infant disappearance at the time of male takeovers accounted for 40
percent of the male and unsexed infant deaths. All infants of primiparous
mothers died (all the infants were male). Among howler females, rank decreases with age, and all the primiparous mothers except one were top ranking.

AOTUS INFULATUS
(OWL MONKEY, NIGHT MONKEY, DOUROCOULIS)

Taxonomy and Distribution

The taxonomy of *Aotus* is debated and remains under study. At one time, all owl
monkeys were grouped as one species, *Aotus trivirgatus* (see Mittermeier and
Coimbra-Filho 1981). However, more recently a number of species are being recognized based on physical appearance, distribution, chromosomal data, and blood
protein variation, with a general division into a northern "gray-necked" group and
a southern "red-necked" group (Ford 1994, Hershkovitz 1983). Hershkovitz (1983)
recognized nine species and eleven subspecies, while Ford (1994) recognizes only
five to seven species. The species in the Guajá area is the red-necked, *Aotus infulatus*, which, according to Ford (1994), may not be distinct from *Aotus azarae*.

Owl monkeys can be found in forests from western Panama south to northern Argentina (Aquino and Encarnación 1994, Wright 1984). They are found in
the Amazon Basin, except north of the Rio Amazonas and east of the Rio Negro,
and they are also absent in the Guianas (Emmons and Feer 1997). *Aotus* inhab-
its the primary, secondary, and remnant tropical forests.

Physical Appearance

The owl monkeys are of small to medium size, weighing approximately 1 kg,
and lacking sexual dimorphism (Wright 1984). The intermembral index is 74

(Napier and Napier 1985). Wright (1981) has further described general physical characteristics of *Aotus*. These monkeys have relatively large eyes and distinctive facial hair coloration. The face has black stripes, and above the eyes are two white or buffy semilunes. The pelage varies in color from brown to gray to reddish with the underside varying from orange to yellow to buff-white. Infants do not exhibit a coat color distinctive from adults (Wright 1984).

Although they are nocturnal animals, owl monkeys lack the tapetum lucidum (Hershkovitz 1983). Wright (1978) and Thorington, Muckenhirn, and Montgomery (1976), indicated that owl monkeys can be located in the forest by their eye reflection. Since owl monkeys lack the true eye-shine of other nocturnal animals, this may aid in distinguishing owl monkeys from other animals in the forest. The eyes of animals with a tapetum lucidum will reflect bright orange when light is shined on them, while the owl monkey reflects pink (Hershkovitz 1983).

Habitat, Locomotion, and Diet

The owl monkey is the only nocturnal New World primate (Wright 1984). Although nocturnal, they are also sometimes seen in the day near their sleeping areas (Hershkovitz 1983). They are able to live in a variety of habitats, such as cloud forests, primary tropical moist forests, secondary forest, remnant tropical forest, and the subtropical dry forests of the Chaco (Aquino and Encarnación 1994; Wright 1984). A wide range of forest strata are used, from 7 to 35 meters (Wright 1981). They sleep in tree hollows or in clumps of tangled vines (Hershkovitz 1983). Locomotion is mainly quadrupedal, with some leaping between trees and with the tail used as a balancing device (Napier and Napier 1985). The diet consists primarily of fruits, which are supplemented with insects, leaves and flowers (Wright 1984).

Activity Patterns and Range

Wright (1978) observed a group of owl monkeys in Peru. Their activity budgets broke down as follows: 21 percent of the time spent travelling, 53 percent of the time spent feeding, 22 percent of the time spent resting, and 4 percent of the time spent in agonistic activity. They traveled an average of 252 meters a night. Their home range was 3.1 hectares, and it overlapped with other groups. Thorington, Muckenhirn, and Montgomery (1976) collected data from nine days of radiotracking an owl monkey released into an unfamiliar forest on Barro Colorado Island. They found that 85 percent of its time was spent in an area of one-half hectare.

Social Organization and Social Behavior

Wright (1984) has described *Aotus* social organization and behavior. The owl monkey lives in monogamous groups composed of an adult pair and one to three offspring. Allogrooming is rare. The adult male and the adult female of the group have different roles with respect to the infant. The adult female nurses the infants and is often the leader of group movements, while the adult male carries, plays with, and shares food with the infant. A study of carrying behavior quantified these differences. During the first week of life, the infant was carried by the father 51 percent of the time, by the mother carried 33 percent of the time, and by a juvenile 15 percent of the time. The juvenile's participation in this decreased as the infant grew larger. By the time the infant was thirty days old, the juvenile did not carry it at all. The father shared food with the infant 27 percent of the time, the mother shared 10 percent of the time, and the juvenile shared 23 percent of the time. Fathers and juveniles also played with infants more often than did the mothers. Juveniles leave the natal group at about two to two and a half years of age.

Intergroup confrontations occur in the wild, and males and females participate equally (Wright 1984). These conflicts occur when other groups attempt to enter an occupied feeding tree (Wright 1978). The owl monkeys vocalize and exhibit threat displays when in confrontation with another group (Wright 1978). Wright (1984) described this specific vocalization as a resonant "war whoop." This consists of a series of 10–17 low notes produced by expansion of the laryngeal sac. Interestingly, Wright (1984) found that in captivity, adults fight with members of the same sex only. Possible evidence of territory advertisement comes from Wright's (1984) observation of an adult male who hooted along the home range borders for 1–2 hours in the nearly full moon an average of once a month.

Reproduction

Studies in captivity indicate that owl monkeys are not seasonal breeders (Hunter et al. 1979). Mating between monkeys kept in pairs has been observed infrequently but continues into pregnancy (Hunter et al. 1979). No evidence has been found for a female postpartum estrus (Hunter et al. 1979). In captivity, the minimum interbirth interval was 148 days, the mean was 271 days, and the median was 248 days (Hunter et al. 1979). The gestation period was 133 days, but this information came from only one recorded pregnancy (Hunter et al. 1979:171). Only one infant is born a year (Wright 1984).

CEBUS APELLA (TUFTED CAPUCHIN, BLACK-CAPPED CAPUCHIN, BROWN CAPUCHIN, COMMON CAPUCHIN) AND CEBUS KAAPORI (KA'APOR CAPUCHIN)

Taxonomy and Distribution

Generally, researchers distinguish four species of *Cebus* spp. (Mittermeier and Coimbra-Filho 1981). These are grouped into an untufted group (*C. albifrons, C. capucinus,* and *C. nigrivittatus*) and a tufted group (*C. apella*). "Tufted" refers to the presence of long, dark, and erect hairs on the top of its head (Freese and Oppenheimer 1981). Some debate exists as to whether members of the tufted group are sufficiently different to warrant separate speciation. *Cebus apella* can be found east of the Andes from Colombia and Venezuela south to Paraguay and Northern Argentina (Emmons and Feer 1997; Wolfheim 1983). *C. nigrivittatus* also goes by the designation *C. olivaceus.*

The Ka'apor capuchin was first identified and described by Queiroz (1992). Queiroz (1992) has described *Cebus kaapori* as an untufted form that bears affinities to *C. nigrivittatus* (*C. olivaceus*) but is longer bodied, more gracile, and separated by more than 400 km from *C. nigrivittatus*. Since his initial description, it has been argued that it is best considered a subspecies of *C. nigrivittatus* (Masterson 1995). Emmons and Feer (1997) classified it as a subspecies of *C. nigrivattatus*; however, Kinzey (1997) supported Queiroz's original classification. Queiroz (1992) found the Ka'apor capuchin in the vicinity of the Guajá reserves; however, it seems to have a patchy distribution as far west as the Tocantins (Ferrari and de Souza 1994; Ferrari and Lopes 1996; Lopes and Ferrari 1996). Lopes and Ferrari (1996:321) describe *Cebus kaapori* as "one of the most, if not *the* most, endangered of Amazonian mammals due to its low population density, restricted geographical range, and threats due to habitat destruction and hunting pressure."

Physical Appearance

Capuchins are medium-sized New World monkeys. *Cebus apella* has an average adult mass of 3,450 grams (Fleagle and Mittermeier 1980). All *Cebus* monkeys have prehensile tails and are sexually dimorphic (Freese and Oppenheimer 1981; Fleagle 1999). *C. apella* has a suite of cranial-mandibular features that allow high degrees of masticatory stress for exploitation of hard nuts and fruits (Cole 1992; Janson and Boinski 1992).

Habitat, Locomotion, and Diet

Fleagle and Mittermeier (1980) and Mittermeier and van Roosmalen (1981) observed *Cebus apella* in the Raleighvallen-Voltzberg Nature Reserve in Central Surinam. *C. apella* was found most frequently in the high forest and liane forest, where it showed a preference for the middle to lower levels of the main canopy. However, they were also found in all other forest types in the reserve: high rain forest, low rain forest, mountain savanna forest, liane forest, and pina swamp forest. They also entered edge habitats. *Cebus apella* is basically a quadrupedal walker (84 percent), with 10 percent leaping and 5 percent climbing during travel. During feeding, it spends 88 percent of the time in quadrupedal locomotion, 4 percent leaping, and 8 percent climbing. It moved primarily on branches and twigs.

Cebus apella also heavily exploits palms, and Terborgh (1983:82) stated that palms "represent the primary ecological refuge of the species." He reported that *Cebus apella* exploits palms for insect foraging, the exocarp or mesocarp of seeds, the apical meristem, and the immature inflorescences. During the dry season, he found that palms and figs represented 97 percent of the feeding time of *Cebus apella* as well as *Cebus albifrons*. Izawa (1975) has described *C. apella* in the Upper Amazon Basin. Behavioral observations indicated that they were eating fruits, except for one occasion each, when they were observed eating nuts and leaves. Examination of stomach contents revealed primarily fruits and insects. The cracking of the palm fruit (*Astrocaryum chambira*) and *Syagus* nuts have also been observed in *C. apella* (Izawa and Mizuno 1977; Ottoni and Mannu 2001.) In addition, Torres de Assumpção (1981) has also observed the feeding behavior of *Cebus apella* in São Paulo, Brazil. He observed *C. apella* licking the inflorescence of the flowering tree *Mabea fistulifera* and suggested that the capuchin may be a pollinator. The Guajá fed meat to all monkeys, and *Cebus apella* was also observed to break bones and eat the softer spongy material within.

Cebus nigrivattatus, which may be related to *C. kaapori*, was sympatric with *C. apella* in Central Surinam but largely restricted to the understory and lower to middle levels of high forest (Fleagle and Mittermeier 1980; Mittermeier and Roosmalen 1981). Observed dietary items of *C. nigrivattatus* were insects, fruit, flowers, leaves, and seeds (Fleagle and Mittermeier 1980; Mittermeier and Roosmalen 1981). Among the Guajá, *C. kaapori* was observed to eat insects as well as meat when provided. Dietary information is available for other *Cebus* species. An analysis of the stomach contents of *Cebus albifrons*, which is sympatric with *C. apella* in some areas, revealed primarily fruits and insects (Izawa

1975). The diet of *C. capucinus* has been described on Barro Colorado Island (Hladik 1978). They obtained a large part of their dietary protein from insects and small invertebrates. The diet consisted of 15 percent leaves, shoots, and other green vegetal material, 65 percent fruit, and 20 percent vertebrate and invertebrate prey.

Activity Patterns and Range

Terborgh (1983) reported the activity pattern for *C. apella* at Coca Cashu, Peru as 66 percent foraging, 21 percent travel, and 12 percent rest. Similar findings were found by Freese and Oppenheimer (1981), with 67 percent foraging, 25 percent travel and 8 percent rest. Terborgh (1983) described the home range of *C. apella* as 80 km. Information on day range is available for *C. capucinus* in Santa Rosa, which was estimated at 2 km (Freese and Oppenheimer 1981). Lopes and Ferrari (1996) observed a group Ka'apor capuchins foraging and resting with a group of bearded sakis.

The capuchins are also noted for their ability to manipulate objects. Numerous studies have documented tool use in captivity and in the wild, particularly in *C. apella*, whose frequency of use arguably rivals that of the chimpanzees (e.g., Fragaszy and Visalberghi 1989; Panger 1998, 1999; Visalberghi and McGrew 1997; Visalberghi and Trinca 1989; Westergaard and Suomi 1997a, 1997b; see also Visalberghi and Fragaszy 1990). Some of the behaviors observed in capuchins include using sticks as probing tools, using sponging tools, using rocks to pound open hard shelled nuts, and using rocks to modify bones to create probing tools. Among the Guajá, brown capuchins were frequently observed to display object manipulation and pounding behavior, particularly of babassu palm nuts. On one occasion, an adult brown capuchin was observed to place a bone on a piece of wood and break it open using a babassu nut as a hammer, so that he could eat the spongy bone within it.

Social Organization and Social Behavior

Izawa (1976) has studied the social organization of the sympatric *C. apella* and *C. albifrons* in the Upper Amazon Basin. *C. apella* temporarily formed small and varied groups that included an adult male and female and her offspring, adult males only, or females with juveniles or infants. When travelling over long distances, one large group of 18–20 individuals moved together. *Cebus albifrons* was observed in groups of adult males and adult females with all the monkeys travelling together as a unit. One group documented by Izawa consisted of 15–20 individuals, and another group encountered only once appeared to have about 50 individuals. On Barro Colorado Island, neighboring *Cebus capucinus* groups have frequent contact with one another, which involve aggressive displays but no loud territorial calls (Hladik 1978).

Dobroruka (1972) has described social communication in captive *C. apella*. Both males and females scent-marked their cages with urine. Sternal epigastric marking was seen in the dominant male. Raised eyebrows indicated a threat. Flattened ears with the corners of the mouth pulled down to expose the canines indicated a defensive posture. A branch shaking display was observed in dominant males or in lower ranking males when the alpha male was at some distance. Tail wagging was seen when excited.

Bernstein (1966) found evidence of the control role held by a male in a group of captive *C. albifrons* in the absence of a dominance hierarchy. The behaviors associated with this were assuming a position between an external disturbance and the group, attacking whatever appears to be distressing a captured group member, and approaching and terminating most cases of intragroup disturbance. Bernstein prefers the term "control animal" to that of "dominant male."

Monkeys in Guajá captivity were typically infants and juveniles, separated from conspecifics and offering little opportunity to observe natural social behavior. However, two visual displays described for *C. nigrivattatus* in Venezuela by Oppenheimer and Oppenheimer (1973) bore similarity to displays seen in the Ka'apor capuchin, which may support a relationship between the two species. These are the display posture involving body orientation toward and staring at another, and the OMBT display (or open-mouth bared-teeth face). These two visual displays were frequently observed in two adult females housed together in a small, three-sided thatch dwelling outside a Guajá hut. The display posture and the OMBT display were often performed simultaneously by the capuchins. The social context involved the approach of humans, either a Guajá or myself, and could conceivably have been either a threat or a submissive gesture. Since the monkeys were tied, they could neither approach nor flee to follow up on the display, so the meaning was difficult to interpret. They were not observed directing these displays towards each other.

Reproduction

Freese and Oppenheimer (1981) have provided some information on reproduction in *C. apella*. In Brazil they have a seasonal birth pattern. They give birth most often from May to June and October to November. Gestation takes five to six months. Mothers give birth to an infant every one to two years. Females come into estrus every eighteen days.

CHIROPOTES SATANAS SATANAS
(BLACK-BEARDED SAKI, SOUTHERN BEARDED SAKI)

Taxonomy and Distribution

Two species of apparently allopatric *Chiropotes* species are commonly recognized: *C. albinasus* and *C. satanas* (Emmons and Feer 1997; Wolfheim 1983). *Chiropotes satanas* are found in the eastern Amazon Basin in Venezuela, Brazil, and the Guianas. They are also found to the north of the Amazon east of the Rios Negro and Orinoco and to the south of the Amazon east of the Rio Xingú. *C. satanas* has been divided into two subspecies: *C. s. satanas* and *C. s. chiropotes*: (van Roosmalen et al. 1981). *C. s. satanas* is found between the Xingu and eastern Amazonia (Johns and Ayres 1987). Prior to the description of the Ka'apor capuchin, Johns and Ayres described *C. satanas satanas* as the most endangered primate in Amazonia.

Physical Appearance

Van Roosmalen, Mittermeier, and Milton (1981) and Fleagle (1988) have described the general physical characteristics of *Chiropotes satanas*. They are medium-sized, weighing approximately 3 kg. They are distinctive in having a black beard, a thick and bushy (non-prehensile) tail, and what has been described as a "bouffant hairdo." No apparent sexual dimorphism exists for either size or color. Both subspecies have a black coat, but *C. s. chiropotes* has a yellowish brown to ochre-colored area from the shoulders to the base of the tail, and *C. s. satanas* has a dark brown to black area on the shoulders and back. *Chiropotes* spp. are known for their extremely large canines, laterally splayed, which are used to break open hard fruits and husks; however, they specialize on the seeds within these plants (Kinzey 1992).

Habitat, Locomotion, and Diet

A series of studies has been done on *Chiropotes satanas chiropotes* in the Raleighvallen-Voltzberg Nature Reserve in Central Surinam (Fleagle and Mittermeier 1980; Mittermeier and van Roosmalen 1981; van Roosmalen, Mittermeier, and Fleagle, 1988; van Roosmalen, Mittermeier, and Milton 1981). The bearded saki was found almost exclusively in the high forest and mountain savannah forest, in the upper part of the canopy, and in emergents. During travel, quadrupedal walking and running accounted for almost 80 percent of the locomotion, leaping accounted for 18 percent, and climbing was rare. During feeding, quadrupedal running accounted for 88 percent of the locomotion, leaping accounted for 9 percent, and climbing accounted for 2 percent. Quadrupedal locomotion took place mainly on relatively large branches, and leaping took place on somewhat smaller supports. Postural behavior was typically above the branches, but some suspensory behaviors (particularly hind limb suspension) were also used when feeding.

C. s. *chiropotes* was observed eating from eighty-five different plant species. They fed primarily on immature seeds (62.2 percent of time) and ripe fruit (30 percent). Leaf stalks (3.4 percent) and flowers were eaten in very small amounts. Norconk, Wertis, and Kinzey (1997) suggested that *C. satanas chiropotes* in Venezuela fills a similar seed predator niche to that of the green macaw (*Ara chloroptera*). Their analysis of stomach contents of four monkeys revealed insects from six orders, but seeds and fruit remained the most abundant. Ayres and Nessimian (1982) also examined stomach contents of *Chiropotes satanas* from Pará, Brazil, along the Nhamund River. They found insect material from eight taxonomic orders. Among the Guajá, one saki, who was at times allowed to move freely in a hut, was observed systematically capturing insects living in the thatching of the hut. In another hut, the Guajá hung a tortoise carapace from a string next to their saki; the saki broke off pieces and consumed them. Sakis living among the Guajá also ate meat when it was provided.

Activity Patterns and Range

Some information is available on activity patterns, day ranges, and home ranges for *C. s. chiropotes* in the Voltzberg study area in Surinam (van Roosmalen, Mittermeier, and Milton 1981). The foraging behavior has been described as rapid movement between food trees with short and intense feeding bouts. The estimated home ranges were very large, from 200–250 hectares. The estimated day ranges were approximately 2,500 meters.

Social Organization and Social Behavior

C. s. *chiropotes* in Surinam were found in multi-male groups, which may have thirty or more members (van Roosmalen, Mittermeier, and Milton 1981). They have a fission-fusion social structure wherein the large group breaks up into small groups to forage during the day (van Roosmalen, Mittermeier, and Fleagle 1988). Only a few social behaviors have been noted among *C. s. chiropotes* (van Roosmalen, Mittermeier, and Milton 1981). Tail-wagging is an apparent indication of excitement. They also have a distinctive, high-pitched whistle which serves as a contact call and when performed more intensely serves as an alarm. Grooming has been occasionally observed. Among the Guajá, taking turns with the high pitch contact call was heard among juvenile *C. s. satanas* living in different huts.

Reproduction

Seasonal births were seen among *C. s. chiropotes* in Surinam (van Roosmalen, Mittermeier, and Milton 1981). The birth season was in December and January with single births being the rule. No evidence of paternal care has been seen in *Chiropotes* (Fleagle 1988). One female saki among the Guajá was observed in estrus with a genital swelling and slight redness. She frequently performed contact calls.

SAIMIRI SCIUREUS (SQUIRREL MONKEY)

Taxonomy and Distribution

Mittermeier and Coimbra-Filho (1981) have described the debate that exists as to how many species are in the *Saimiri* genus, which exhibit only moderate variation in coat color and facial features. Hershkovitz (1984) has identified four species which can be grouped into a "Roman" type (*S. boliviensis*) and a "Gothic" type (*Saimiri sciureus, S. oerstedi,* and *S. ustus*), based on the shape of the circumocular patch. The species in the Guajá area, by any classification scheme, is *Saimiri sciureus* and is anatomically of the Gothic type (Baldwin and

Baldwin 1981). Baldwin and Baldwin describe the distribution based on a two-species classification. *Saimiri sciureus* can be found in Amazonian Brazil, Colombia, Ecuador, Peru, Bolivia, the Guianas, and south of the Rio Orinoco in Venezuela. It also is found in parts of Columbia and the Bolivian basin. *Saimiri oerstedi* is found in Costa Rica and Panama.

Physical Appearance

Saimiri sciureus are small primates with an average adult weight of 688 grams (Fleagle and Mittermeier 1980). They are sexually dimorphic with females typically ranging between 500 and 750 grams and males ranging between 700 and 1,100 grams. The muzzle is black, the throat, face and ears are white, and the pelage is relatively short, yellow or greenish-yellow in color, with white or light yellow undersides (Baldwin and Baldwin 1972).

Habitat, Locomotion, and Diet

Fleagle and Mittermeier (1980) and Mittermeier and van Roosmalen (1981) observed the habitat, locomotion pattern, and diet of *Saimiri sciureus* in the Raleighvallen-Voltzberg Nature Reserve in Surinam. The squirrel monkeys were found in all types of forest, but the liane forest was preferred. While other monkeys in the area were rarely found in the liane forest, the squirrel monkeys were found there more than 50 percent of the time, primarily in the forest understory. The squirrel monkey locomotion pattern was predominantly quadrupedal during travel (87 percent), with some leaping (11 percent) and a small amount of climbing (2 percent) (Fleagle and Mittermeier 1980).

Fleagle and Mittermeier (1980) observed dietary consumption of insects, fruit, and flowers, but not leaf eating or seed eating. Other researchers have also made observations on *Saimiri* diets. Izawa (1975) observed *Saimiri sciureus* in the basin of the Caqueta River in the Upper Amazon, where they were seen to eat fruit and nuts. Analysis of stomach contents indicated the monkeys were also eating insects. Fragaszy (1978) noted that in captivity, *Saimiri sciureus* grab food and eat it quickly, which may be an adaptation to eating insects. Among the Guajá, a female squirrel monkey, who was rarely tied, was observed on several occasions eating grasshoppers near Guajá huts.

Activity Patterns and Range

Baldwin and Baldwin (1972) observed *Saimiri oerstedi* in their natural habitat in a forest in Panama. The monkeys became active at daybreak, spending 95 percent of each hour of their 14-hour day travelling and foraging. The day range was between 2.5 and 4.2 km per day. Their home range was 0.175 km² (43.5 acres). Terborgh (1983) described the home range of *S. sciureus* at Cocha Cashu, Peru, as greater than 250 hectares.

Social Organization and Social Behavior

Baldwin (1968, 1971) made observations on squirrel monkey social organization on a troop of over one hundred semi–free ranging *Saimiri sciureus* in Monkey Jungle, a semi-natural environment in South Florida. The troop typically traveled as a unit during the day but divided into subgroups at night for sleeping. Sometimes adult and subadult males traveled separately. Most animals tended to interact affiliatively and travel near members of the same age and sex. The exception was adult males in the mating season. A dominance hierarchy was observed among males during the mating season, but dominance did not involve control of troop movements. Males did, however, investigate the source of alarm calls. No dominance hierarchy was seen among the females, but females controlled the group's movements. The interactions between adult males and adult females often led to threats and chases, and rarely included affiliative interactions.

Similar findings were made by Baldwin and Baldwin (1972) among *Saimiri oerstedi* in a natural habitat in Panama. Adult females, infants, juveniles, and subadult males traveled as a cohesive, integrated unit. The troop's two adult males usually traveled at the edge of the troop, with a young-adult male travelling even further at the periphery of the troop. Troop size was smaller than in Monkey Jungle, with only twenty-three members. All overt social interactions were infrequent. Home ranges were not defended.

Izawa (1976) observed *S. sciureus* in the Upper Amazon Basin and also found similar patterns of social organization. *S. sciureus* was seen in large groups with thirty-five to fifty individuals. They separated into smaller temporary groups during the day. Both adult males and adult females are found in a group. When moving together in a file, the monkeys tended to group according to age and sex. Sets of adult males were often at the end of the file. Mason (1978) also found that squirrel monkeys tended to interact more frequently with individuals of the same sex and age rather than engaging in heterosexual pair-bonding.

At Monkey Jungle, clear differences in social behavior occurred during the mating season (Baldwin 1968, Baldwin and Baldwin 1972). Males, who were most often socially inactive and nonaggressive, became excitable, aggressive, and highly vocal. The dominance hierarchy was expressed overtly through penile displays. Stereotyped urine-washing and urine-wash-kicking were also seen among adult males during the mating season.

In captivity, it has been noted that *S. sciureus* exhibit a marked increase in vocalizations at nightfall (Symmes and Goedeking 1988). An increase in this vocal behavior occurred when investigators produced an abrupt loss of visual contact among monkeys. The researchers interpreted this behavior as functioning to maintain social cohesion, which outweighs the risk of drawing predator attention. Winter (1969b) distinguished several types of squirrel monkey calls. Broadly, these were peep calls (alarm peep, isolation peep, play peep, peep in other contexts) and twit calls (trill, twitter, twit-like, vit, and vit with hook calls). All but play peeping were distinguishable by either the parameter of frequency or duration, or both. Play peeping can be mistaken for vit calls by pectrographic analysis; however, in context of the use of the call is clear.

Reproduction

Du Mond and Hutchinson (1967) have observed the reproductive behavior of *Saimiri sciureus* at Monkey Jungle. They are seasonal breeders, with a mating season in Florida from mid-December through March. Adult males display seasonal secondary sex characteristics associated with seasonal spermatogenesis. This is termed the "fatted" state. The pelage becomes thicker and fluffier especially around the upper torso, shoulders, and arms. Weight increases from an average of 715 grams (non-fatted) to 800 grams (fatted). However, among a troop of *Saimiri oerstedi* in a natural habitat in Panama, the "fatted" state was not seen in the mating season (Baldwin and Baldwin 1972). The Baldwins interpreted this as likely being related to food availability.

SAGUINUS MIDAS NIGER
(RED-HANDED TAMARIN, MIDAS TAMARIN)

Taxonomy and Distribution

According to Mittermeier and Coimbra-Filho (1981), the taxonomy of the tamarins established by Hershkovitz (1977) has not been seriously criticized.

Ten species are identified, which are divided into three basic groups: "bare-faced" (e.g., *S. bicolor* and *S. oedipus*), "mottled-faced" (e.g. *S. inustus*), and the "hairy faced" tamarins (white-mouth tamarins, the moustached tamarins, and the Midas tamarins). The white-mouthed tamarins include *S. nigricollis* and *S. fuscicollis*; the moustached tamarins include *S. mystax*, *S. labiatus*, and *S. imperator*; and the Midas tamarin (*S. midas*) includes two subspecies: *S. midas midas* and *S. midas niger*. *S. midas* can be found in Brazil, the Guianas, and the eastern Amazon Basin east of the Rio Negro and the lower Rio Xingú (Emmons and Feer 1997). *S. midas midas* is found above the Amazon river and *S. midas niger* is found in areas below the Amazon river, including Marajó Island (Hershkovitz 1977), which includes the Guajá reserves.

Physical Appearance

Saguinus midas are small monkeys but medium-sized among tamarins. Weights from three male *S. midas midas* averaged 533 grams, while a female weighed 450 grams (Fleagle and Mittermeier 1980). The *S. midas niger* pelage is black with yellowish mottling on the back but without the orange colored feet of *S. m. midas* (Hershkovitz 1977).

Habitat, Locomotion, and Diet

Fleagle and Mittermeier (1980) and Mittermeier and van Roosmalen (1981) observed *Saguinus midas midas* in the Raleighvallen-Voltzberg Nature Reserve in Central Surinam. *S. m. midas* was found mainly in the understory and lower to middle parts of the canopy of all five forest types in the study area: high rain forest, low rain forest, mountain savanna forest, liane forest, and pina swamp forest. It spent more time in edge than non-edge habitats and a preference was also seen for high forest. *S. m. midas* utilized primarily quadrupedal walking, running, and bounding (accounted for 76 percent while travelling). They also utilized some leaping locomotion (accounted for 24 percent while travelling). Locomotion during feeding involved more quadrupedal movement (87 percent), with some leaping (12 percent), and occasional climbing (2 percent). The monkeys were observed eating a variety of plant foods (fruit, flowers, leaves, buds, and tree exudates) and to a lesser extent, insects. Kessler (1995) observed an increase in fruit eating at the beginning of the wet season for *S. m. midas* in the wet season, and insect eating in the dry season seemed to influence the way it used its home range in the dry season. In contrast to Fleagle and Mittermeier (1980), Kessler observed a distinct preference for edge habitats.

Activity Patterns and Range

The activity pattern described for *S. m. niger* is 42 percent locomotion, 19 percent affiliative social interaction, 19 percent personal care, 12 percent feeding, and 8 percent aggression (Hershkovitz 1977). Locomotion and feeding occurred most often in midday. The home range for *S. midas* has been measured at 9 hectares (Snowdon and Soini 1988).

Social Organization and Social Behavior

Snowdon and Soini (1988) described aspects of tamarin social organization. The core of tamarin groups consisted of one reproducing female, her mate, and her offspring. Other adults or subadults may be attached to the group. Among *S. midas*, group size has been observed to range from three to six individuals.

Vogt, Carlson, and Menzel (1978) and Vogt (1978) observed the social behavior of *S. fuscicollis*. Observations were made of a group saddle-back tamarins over a thirty-two-month period in a seminatural environment. During this time, three sets of infants were born. During the first two weeks of life, infants were carried predominately by fathers. During the third and fourth weeks, no significant difference was found in the amount of time the infant was carried by the father, the mother, and non-parents. After the fourth week, the father and non-parents carried the infant about equally, with the mother carrying very little. With respect to sex, the males carried infants more often that females, except during the third and fourth weeks when they were about equal. Further observations were made of a group of saddleback tamarins as group size changed between four and seven members over a period of 70 weeks. The adult male and female were usually spatially independent of each other, except during the female's estrus and during caretaking of very young infants. Infants were usually spatially tied to the father until they shifted to subgroups of juveniles and subadults. As the monkeys neared adulthood, they became more spatially independent.

Reproduction

French, Abbott, and Snowdon (1984) have described accumulating evidence for the social environment affecting reproductive hormone levels among tamarins. They observed five female *Saguinus oedipus* tamarins and measured urine and estrogen excretion levels in two social environments. Females were studied in a group with a dominant cycling female and also studied while housed with an unrelated male. Females housed with a dominant cycling female demonstrated low levels of estrogen, did not have cyclic patterns of estrogen secretion, and had low rates of anogenital marking, sexual interactions, and reception of anogenital sniffing from males. When housed with an unrelated male, estrogen levels increased, as did rates of sexual interactions, anogenital marking, and reception of anogenital sniffing from males. In addition, three females began showing cyclic patterns of estrogen excretion when housed with an unrelated male.

NOTES

Introduction

1. The common name for *Alouatta* that has been in popular usage is "howler monkey." Due to the grammatical problem of the double noun, two alternatives are now being used, "howling monkey" or, more simply, "howler." In this work, the latter is used.

1. A Brief History of the Guajá

1. Lange (1914) also described the Amanayé to the east.
2. Cormier (in press) provides an abbreviated description of Guajá recent history.

2. A Brief History of New World Monkeys

1. Physical differences among members of the *Saimiri* genus are moderate and debate exists as to how many species should be taxonomically distinguished within the genus (Mittermeier and Coimbra-Filho 1981).

2. Macaques have also been used in aerospace research (e.g., Wiebers et al. 1997).

3. One example of these practices from outside of Amazonia is found in Sprague's (2002) reports that monkeys were highly prized for their supposed medicinal properties prior to WWII.

4. The possibility exists that the respiratory infections observed in the monkeys are due to simian diseases. However, the reports suggest anthroponosis since the monkeys died from the respiratory infection.

5. An abbreviated description of primate adaptation to anthropogenic forests appears in Cormier (2002b).

6. The Ka'apor may have a patchy distribution further west, as indicated by Lopes and Ferrari's (1996) identification of a Ka'apor capuchin killed by a hunter near the Tocantins river to the west.

3. Monkey Hunting

1. Lowland South American Indians whose use of monkeys as food has been studied include: the Aché (Hill and Hawkes 1983; Kaplan and Kopischke 1992), the Aguaruna (Berlin and Berlin 1983), the Araweté (Viveiros de Castro 1992), the Barí (Lizarralde 1997), the Bororo (Crocker 1985), the Caqueta River Indians (Izawa 1975), the Camayura (Meggers 1971), the Cashinahua (Kensinger et al. 1975), the Desana (McDonald 1977), the Huambisa (Berlin and Berlin 1983), the Huaorani (Yost and Kelley 1983), the Jívaro (McDonald 1977; Meggers 1971), the Ka'apor (Balée 1984), the Kagwahív (Kracke 1978), the Kalapalo (Basso 1973), the Kayapo (McDonald 1977), the Machiguena (Kaplan and Kopischke 1992; Shepard 1997), the Maku (Milton and Nessimian 1984), the Makuna (Århem 1981), the Mehinaku (Gregor 1977), the Mekranoti (Werner 1984), the Mundurucú (R. F. Murphy 1960), the Pakaa Nova (or the Wari') (Conklin 2001; Von Graeve 1989), the Shiriana (Montgomery 1970), the Siona-Secoya (Vickers 1988), the Sirionó (McDonald 1977; Holmberg 1985 [1950]), the Suyá (Seeger 1981), the Tapirapé (McDonald 1977; Wagley 1983 [1977]), the Trumaí (Murphy and Quain 1966), the Xavante (Maybury-Lewis 1967; McDonald 1977), and the Yanomamö (McDonald 1977, Smole 1976).

2. A number of Amazonian groups eat insects, including the Aché (see Hill and Hawkes 1983; Hurtado et. al 1985; Métraux and Baldus 1946), the Cubeo (Goldman 1948), the Ka'apor (Balée 1988), the Tukano (Dufour 1987), and the Yanomamö (Lizot 1977; Smole 1976; also see Gilmore 1950), as well as non-Amazonian peoples (see Bodenheimer 1951; Defoliart 1988; Ramos-Elorduy 1988; Ratcliffe 1988; Sutton 1988).

3. An abbreviated discussion of the ecological role of monkeys appears in Cormier (2002a).

4. Species were identified through various zoological references to local animal life (e.g., Cristina dos Santos et al. 1995; Emmons 1990, 1997; Gery 1977; Goulding

1980; Kricher 1989; Mascarenhas, de Fátima Cunha Lima, and Overal 1992; Meyer de Schauensee 1964; Queiroz 1992).

5. Forline (1997) did use this method during his research with the Guajá, using individuals although he does note similar food-sharing behaviors. The reader is referred to the nutritional analyses in his study.

6. A similar practice has been noted among the linguistically related Kagwahiv (Parintintin). Kracke (1978:21) described their principle of obligatory food sharing as "ask and you shall be given," where it is considered extremely rude to fail to share food with anyone present.

7. Queiroz and Kipnis (1991) found that monkeys were the most utilized animal type on the Caru reserve through excavation and analysis of animal bone remains from an abandoned Guajá camp. Excluding fish, they estimated that 32.5 percent by weight of game animals were primates with 26 percent coming specifically from howlers. Forline (1997) analyzed game weights on three different Guajá indigenous posts. Although his data differs somewhat, it generally conforms in that monkeys are among the top three species exploited in the wet season and fish are among the top three species exploited in the dry season, in terms of dressed weight.

8. In evaluating the seasonal differences, an adjustment was made to correct for the differences in the number of observations in the late wet season and the early dry season. In the late wet season, there were 61 sampling days and 3,020 observations. In the early dry season there were 50 sampling days and 2,186 observations. Therefore, the early dry-season observations were multiplied by 1.38 to correct for the lesser number of observations overall during this time period.

9. The Portuguese term *curimatá* includes both prochilodins and curimatins. One of the key distinguishing features between these two similar genera is that the prochilodins have a sucking disk at their mouths which the curimatins lack. The sucking disk is characteristic of the fish the Guajá call *piračũ* (see Gery 1977 and N. Smith 1981).

10. Forline (1997) suspects that fish poisons may be a recent introduction, since the Guajá at Juriti had no knowledge of fish poisons. As such, this may provide additional evidence of cultural loss among the Guajá.

11. According to Forline (1997), Guajá informants indicated that night hunting in general was a hunting strategy introduced to them by the FUNAI.

12. While no special properties of fish could be ascertained, it is notable that Rival (1993) found that the Huaorani do believe that eating certain foods and the application of certain plants promote growth and development during different stages of life.

13. Approximately 75 percent of the plants collected were identifiable, at least to the family level.

14. In other words, 43.64 percent of the known plants were edible by monkeys while 56.36 percent were not.

15. In a somewhat different sense, Shepard (1997) describes monkeys as figuring heavily in Machiguenga ethnobotanical knowledge. He reports that more than fifty species, which is more than a quarter of their medicinal plants, are used as hunting medicines in a culture that focuses heavily on monkey hunting.

16. The Guajá on the Caru most often pronounced the term as *maku*. Nasalization of initial consonant bilabial stops may be in free variation.

17. Such kidnappings and escapes of non-Indians have provided early knowledge of some Amazonian groups such as the Yanomamö (Biocca 1996) and the Tupinambá (Staden 1928 [1557].

18 The Yuqui and the Sirionó also lack fire-making ability (see Balée 1999)

4. Guajá Kinship

1. Godelier, Trautmann, and Tjon Sie Fat (1998) edited a fascinating volume, *Transformations of Kinship*, which focuses on Dravidian and Iroquois kinship systems in cross-cultural perspective. In this work, I am only touching on a few ideas of some of the contributors to that volume.

2. Erikson (in press [MS. 15]) prefers the term polyandrous conception because the term partible paternity, "seems to imply that paternity is fundamentally one, its basic unity only being accidentally transgressed (hence 'partible')."

3. Some groups do believe that there is a role for females, although they share the belief that semen is the essential fetal ingredient. For example, the Wari' (Pakaa Nova) believe that menstrual blood also has a role in conception (Vilaça 1992:52)

4. Both *mapara* and marriage partners are classificatory spouses.

5. Erikson also describes premarital avuncular sexual relations as occurring between a betrothed child and her mother's brother (future father-in-law) among the Matis.

5. Animism and the Forest Siblings

1. Bird-David's work focuses outside of Amazonia, but her description of animism among the Nayaka of India addresses themes that are also important to Amazonia. The Nayaka have a notion called *devaru*, which is a supernatural power manifested in some animals, landscape features, and humans undergoing spirit possession. According to Bird-David, Nakaya animism represents a relational epistemology in that *devaru* is not believed to exist in all nonhumans, but rather becomes known by means of social interaction with nonhumans. While Viveiros de Castro (1999a) takes issue with her characterization of animism as an epistemology rather than an ontology, their views (as well as that of Descola) are similar to the extent that they stress the importance of understanding animism as a social relation.

2. According to Forline (1997) the *mihúa* translates as "strange," "savage," or "dirty."

3. Balée (personal communication) suggested that the *parenči* suffix is derived from the Portuguese "parente" meaning relative.

4. The Guajá also refer to their spiritual aspect as *yára*, which is also a suffix used for the names of the masters or controllers of animals

5. Somewhat similarly, Halbmayer (2001) describes a phenomenon of multiple souls among Carib-speakers, which can involve multiple souls in the body, multiple manifestations of the soul outside the body (such as in landscape features), or multiple destinies of souls at death..

6. Kracke (1999) describes a special grammatical form for telling dreams among the Parintintin through one of their past-tense marking particles.

7. It should be noted however that the Guajá are generally reluctant to talk about unpleasant things. Saying they do not have bad dreams may not mean that they do not; it may mean that it is not discussed.

8. The Guajá also use the term *Tapáno* to refer to the Christian creator god. This is due to the use of a Tupian term for the Christian god by priests who have visited the Guajá. The Guajá describe different attributes of the *Tapáno* of the priests (the maker of all things, who actually bears more similarity to Mai'ira) and their indigenous *Tapáno* (the divinity of rain, thunder, and lightning).

9. There does not seem to be an indigenous word for malaria; rather, it is recognized by its symptoms being alternately *haku*, hot from fever, or *hača'ə*, cold with the chills from a fever.

10. Kracke (1999) has noted a similar use of light to ward off ghosts (*añang*) among the Parintintin. Among the Wari', Conklin (2001) describes a form of dead spirit (*jima*) that purportedly avoids the light.

6. Pet Monkeys

1. Breastfeeding monkeys has also been reported among the Yanoama (Smole 1976) and the Huaorani (Rival 1993).

2. Activity budgets are available for *Aotus infulatus* but were not compared, as I observed owl monkeys in the daytime rather than at night.

3. While the research here did not demonstrate monkeys acting as guards, it is interesting to note that Clastres (1998) makes a brief mention of the Guayaki capturing coatis to use as guard animals in their camps.

7. Cosmology and Symbolic Cannibalism

1. Myerhoff and Ruby provide a similar definition of reflexivity as the "capacity of any system of signification to turn back upon itself, to make itself its own object by referring to itself" (1982:2).

2. This type of widespread cultural information may be due to cultural borrowing or to ancient origins, similar to what Balée (2000) has described as TEK (Traditional Ethnobiological Knowledge) among Tupí-Guaraní groups.

3. The Wari' contrastive categories of *wari'* and *karawa* strike me as similar to the Guajá *wari* (howler) and *karawa*. The Wari' distinguish two basic classifications of life forms, Wari' (human) and *karawa* (nonhuman or non-Wari) (Conklin 2001, Vilaça 1992), which, according to Vilaça, also have the meanings of *the eater* and *the eaten*, respectively. The terms are remarkably similar to the Guajá *wari* and *karawa* terms. While the Wari' *wari'* means *human* and the *eater*, it is very similar to the Guajá term *wari*, which means the formerly human howler who is to be eaten. The Wari' use the food-term *karawa*, which in Guajá denotes the divinities who are in the place of the dead, which can possibly be interpreted as well as the place of those already eaten. The significance of the similarity is unclear. The Wari' are Chapacuran speakers, so the terms should not be cognates. The possibility exists that these terms may have been borrowed, perhaps not directly between the Wari' and Guajá but perhaps between other Tupí-Guaraní and Chapacuran speaking groups in the past, or from the Tupian Língua Geral. Other possibilities are that the terms do represent ancient Amazonian Indian links, given the similarity in their meanings. However, the possibility cannot be rejected that this is coincidental independent coinage of similar sounding terms. Further evidence for a possible relation may be found in the history of the term *karaí* in Tupí-Guaraní. According to Shapiro (1987), another common Tupí-Guaraní term for non-Indian Brazilian, *karaí*, at one time meant a shaman who received power from the creator deity or culture hero (Mai'ira). The Guajá, in addition to using the term to denote non-Indian Brazilians, also have a divinity called Karaí. This further supports the structural similarity of *karawa* (divinity) in Guajá and *karawa* (non-Indian Brazilian) in Wari'

Conclusion. Ethnoprimatology In Amazonia and Beyond

1. According the Fouts, Fouts, and Waters (2002) the chimpanzee "Jerome" received a more virulent strain of HIV than normally occurs as a primary in humans, and, secondly, he did not provide a good model because his immune system was weakened by a lifetime of captivity and biomedical experiments

2. The cause of Baby Fae's death was blood incompatibility rather than outright organ rejection. The child was type O and the baboon was AB (Nowak 1994)

REFERENCES

Abee, C. R. 1989. The squirrel monkey in biomedical research. *Institute of Laboratory Animal Research* 31:11–20.

——. 2000. Squirrel monkey (*Saimiri* spp.) research and resources. *Institute for Laboratory Animal Research* 41:2–9.

Agassiz, L. and E. Agassiz. 1868. *A journey in Brazil*. Boston: Ticknor and Fields.

Aldous, P. 2000. Protests force primate farm to close. *Nature* 404:215

Almeida, R. T., D. S. Pimentel, and E. M. Silva. 1995. The red-handed howling monkey in the state of Pernambuco, north-east Brazil. *Neotropical Primates* 3:174–76.

Altmann, J. 1974. Observational study of behavior: Sampling methods. *Behaviour* 49:227–67.

Aquino, R. and F. Encarnación. 1994. Owl monkey populations in Latin America: Field work and conservation. In J. F. Gaer, R. E. Weller, and I. Kakoma, eds., *Aotus: The owl monkey*, pp. 59–95. San Diego: Academic Press.

Arambourg, C. 1955. A recent discovery in human paleontology: Atlanthropus of Ternifine (Algeria). *American Journal of Physical Anthropology* 13:191–202.

Arens, W. 1979. *The man-eating myth: Anthropology and anthropophagy*. New York: Oxford University Press.

Århem, K. 1981. *Makuna social organization: A study in descent, alliance and the formation of corporate groups in the north-western Amazon*. Stockholm, Sweden: Almqvist & Wiksell International.

——. 1996. The cosmic food web. In P. Descola and G. Pálsson, eds., *Nature and society: Anthropological perspectives*, pp. 185–204. New York: Routledge.

Arquivo Público do Estado do Pará. 1864a. Letter to the President of the Province of Pará. *Patentes e principais de Indios. 1863–1888.* Belém, Pará.

——. 1864b. Letter to the President of the Province of Pará. *Patentes e principais de Indios. 1863–1888.* Belém, Pará.

——. 1871. Letter to the President of the Province of Pará. *Patentes e principais de Indios. 1863–1888.* Belém, Pará.

——. 1873. *Cathechese e civilisação de Indios.* Belém, Pará.

——. 1874. *Catecheses e directores de Indios.* Belém, Pará.

——. 1875. Letter to the President of the province of Pará. *Patentes e principais de Indios. 1863–1888.* Belém, Pará.

——. 1877a. Letter to the President of the province of Pará. *Patentes e principais de Indios. 1863–1888.* Belém, Pará.

——. 1877b. Letter to the President of the province of Pará. *Patentes e principais de Indios. 1863–1888.* Belém, Pará.

——. 1878. Letter to the President of the province of Pará. *Patentes e principais de Indios. 1863–1888.* Belém, Pará.

——. 1887. Letter to the President of the province of Pará. *Patentes e principais de Indios. 1863–1888.* Belém, Pará.

Ayres, J. M. and J. L. Nessimian. 1982. Evidence for insectivory in *Chiropotes satanas. Primates* 23:458–59.

Bailey, L. L. and S. R. Gundry. 1997. Survival following orthotopic cardiac xenotransplantation between juvenile baboon recipients and concordant and discordant donor species: Foundation for clinical trials. *World Journal of Surgery* 21:943–50.

Bailey, R. C. and T. N. Headland. 1991. The tropical rain forest: Is it a productive environment for human foragers? *Human Ecology* 19:261–85.

Bailey, R. C. and N. R. Peacock. 1988. Efe Pygmies of northeast Zaïre. In I. de Garine and G. Harrison, eds., *Coping with uncertainty in the food supply,* pp. 88–117. Oxford: Clarendon Press.

Bailey, R. C., G. Head, M. Jenike, B. Own, R. Rechtman, and E. Zechenter. 1989. Hunting and gathering in the tropical rain forest: Is it possible? *American Anthropologist* 91:59–82.

Baldwin, J. D. 1968. The social behavior of adult squirrel monkeys (*Saimiri sciureus*) in a seminatural environment. *Folia Primatologia* 9:281–314.

——. 1971. The social organization of a semifree-ranging troop of squirrel monkeys (*Saimiri sciureus*). *Folia Primatologia* 12:45–61.

——. 1981. The squirrel monkeys, genus *Saimiri.* In A. F. Coimbra-Filho and R. A. Mittermeier, eds., *Ecology and behavior of Neotropical primates.* Vol. 1, pp. 277–330. Rio de Janeiro: Academia Brasileira de Ciências.

Baldwin, J. D. and J. Baldwin. 1972. The ecology and behavior of squirrel monkeys (*Saimiri oerstedi*) in a natural forest in western Panama. *Folia Primatologia* 18:161–84.

——. 1976. Primate populations in Chiriquí, Panama. In R. W. Thorington Jr. and P. B. Heltne, eds., *Neotropical primates: Field studies and conservation,* pp. 20–31. Washington: National Academy of Sciences.

Balée, W. 1984. The persistence of Ka'apor culture. Ph.D. diss., Columbia University. Ann Arbor: Microfilms International.

——. 1985. Ka'apor ritual hunting. *Human Ecology* 13:485–510.

——. 1988. Indigenous adaptation to Amazonian palm forests. *Principes* 32: 47–54.

——. 1989. The culture of Amazonian forests. In D. Posey and W. Balée, eds., *Resource management in Amazonia: Indigenous and folk strategies*. Vol. 7 of *Advances in economic botany*, pp. 1–21. Bronx: New York Botanical Garden.

——. 1992. People of the fallow: a historical ecology of foraging in lowland South America. In K. H. Redford and C. Padoch, eds., *Conservation of Neotropical forests: building on traditional resource use*, pp. 35–57. New York: Columbia University Press.

——. 1994a. *Footprints of the forest: Ka'apor ethnobotany—The historical ecology of plant utilization by an Amazonian people*. New York: Columbia University Press.

——. 1994b. The destruction of pre-Amazonia: governmental negligence versus indigenous peoples in eastern Brazilian Amazonia. Prepared statement read on May 10 before the Subcommittee on Western Hemisphere Affairs, Committee on Foreign Affairs, U.S. House of Representatives.

——. 1996. On the probable loss of plant names in the Guajá language (eastern Amazonian Brazil). In S. K. Jain, ed., *Ethnobiology in human welfare*, pp. 473–81. New Delhi: Deep Publications.

——. 1998. Historical ecology: premises and postulates. In W. Balée, ed., *Advances in historical ecology*, pp. 13–29. New York: Columbia University Press.

——. 1999. Mode of production and ethnobotanical vocabulary: a controlled comparison of Guajá and Ka'apor (eastern Amazonian Brazil). In T. L. Gragson and B. Blount, eds., *Ethnoecology: Knowledge, resources, and rights*, pp. 24–40. Athens: University of Georgia Press.

——. 2000. Antiquity of traditional ethnobiological knowledge in Amazonia: The Tupí-Guaraní family and time. *Ethnohistory* 47:399–422.

Barnes, J. A. 1947. The collection of genealogies. *Rhodes Livingston Journal* 5:52–53.

Basso, E. B. 1973. *The Kalapalo Indians of central Brazil*. New York: Holt, Rinehart, and Winston.

——. 1977. The Kalapalo dietary system. In Ellen B. Basso, ed., *Carib-speaking Indians: Culture, language, and society*, pp. 98–105. Vol. 28 of *Anthropological Papers of the University of Arizona* Tucson: University of Arizona Press.

Beckerman, S. 1983. Carpe diem: an optimal foraging approach to Bari fishing. In R. B. Hames and W. T. Vickers, eds., *Adaptive responses of native Amazonians*, pp. 269–99. New York: Academic Press.

Beghin, F. 1951. Les Guajá. *Revista do Museu Paulista*, N.S. 5:137–39.

——. 1957. Relation du premier contact avec les indiens Guajá. *Journal de la Société des Américanistes*, N.S. XLVI:197–204.

Bennett, C. F. 1968. Human influences on the zoogeography of Panama. *Ibero-Americana* 51:1–112.

Berlin, B. and E. A. Berlin. 1983. Adaptation and ethnozoological classification: Theoretical implications of animal resources and diet of the Aguaruna and Huambisa.

In R. B. Hames and W. T. Vickers, eds., *Adaptive responses of native Amazonians*, pp. 301–25. New York: Academic Press.

Bernstein, I. 1966. Analysis of a key role in a capuchin (*Cebus albifrons*) group. *Tulane Studies in Zoology* 13:49–54.

——. 1976. An appeal for the preservation of habitats in the interests of primate conservation. *Primates* 17:413–17.

Bernstein, I., P. Balcaen, L. Dresdale, H. Gouzoules, M. Kavanaugh, and T. Patterson. 1976. Differential effects of forest degradation on primate populations. *Primates* 17:401–11.

Bettendorf, J. F. 1910. *Crônica da Missão dos Padres da Companhia de Jesus no Estado do Maranhão*. Rio de Janeiro: J. Leite.

Binford, L. S. B. 1977. Olorgesailie deserves more than the usual book review. *Journal of Anthropological Research* 33:493–502.

Biocca, E. 1996. *Yanoama: The Story of Helena Valero, a Girl Kidnapped by Amazonian Indians*, trans. D. Rhodes. New York: Kodansha International.

Bird-David, N. 1999 "Animism" revisited: Personhood, environment, and relational epistemology. *Current Anthropologist* 40:S67–S91.

Bobadilla, U. L. and S. F. Ferrari. 2000. Habitat use by *Chiropotes satanas utahicki* and syntopic platyrrhines in eastern Amazonia. *American Journal of Primatology* 50:215–24.

Bodenheimer, F. S. 1951. *Insects as human food: A chapter in the ecology of man*. The Hague: Junk Publishers.

Bökönyi, S. 1969. Archaeological problems and methods of recognizing animal domestication. In P. J. Ucko and G. W. Dimbleby, eds., *The domestication of plants and animals*, pp. 219–29. Chicago: Aldine.

Bonvicino, C. R. 1989. Ecologia e comportamento de *Alouatta belzebul* (Primates: Cebidae) na mata atlántica. *Revista Nordestina de Biologia* 6:149–79.

Bonvicino, C. R., A. Langguth, and R. Mittermeier. 1989. A study of pelage color and geographic distribution in *Alouatta belzebul* (Primates: Cebidae). *Revista Nordestina de Biologia* 6:139–48.

Bowden, D. M. and O. A. Smith. 1992. Conservationally sound assurance of primate supply and diversity. *Institute for Laboratory Animal Research* 34:53–56.

Bown, T. M. and C. N. Larriestra. 1990. Sedimentary paleoenvironments of fossil platyrrhine localities, Miocene Pinturas Formation, Santa Cruz Province, Argentina. *Journal of Human Evolution* 19:87–119.

Brito, C. M. 1991. Índios das "corporações": Trabalho compulsório no Grão-Pará nos esquemas do directório. Belém, Pará: Arquivo Público do Estado do Pará.

Brosius, J. P. 1991. Foraging in tropical rain forests: The case of the Penan of Sarawak, east Malaysia (Borneo). *Human Ecology* 19:123–50.

Burkill, I. H. 1954. Dioscoreaceae. In C. G. van Steenis, ed., *Flora Malesiana, Series I: Spermatophyta*, pp. 293–347. Jakarta: Nordoff.

Burton, F. D. 2002. Monkey King in China: Basis for a conservation policy? In A. Fuentes and L. Wolfe, eds., *Primates face to face: The conservation implications of human-nonhuman primate interconnections*, pp. 137–62. Cambridge: Cambridge University Press.

Butler, D. 1999. FDA warns on primate xenotransplants. *Nature* 398:549.

———. 2000. U.S. decides close tabs must be kept on xenotransplants. *Nature* 405:606–7.

Caccone, A. and J. R. Powell. 1989. DNA divergence among hominoids. *Evolution* 43:925–42.

Campbell, A. T. 1989. *To Square with Genesis: Causal statements and shamanic ideas in Wayãpí.* Edinburgh: Edinburgh University Press.

Carneiro, R. L. 1964. Shifting cultivation among the Amahuaca of eastern Peru. *Volkerkundliche Abhandlungen* 1:9–18.

Cartelle, C. and W. C. Hartwig. 1996a. A new extinct primate among the Pleistocene megafauna of Bahía, Brazil. *Proceedings of the National Academy of Science of the United States of America* 93:6405–9.

———. 1996b. Updating the two Pleistocene primates from Bahía, Brasil. *Neotropical Primates* 4:46–48.

Carter, A. and C. Carter. 1999. Cultural representations of nonhuman primates. In P. Dolhinow and A. Fuentes, eds., *The nonhuman primates*, pp. 270–76. Mountain View, Calif.: Mayfield.

Cavalieri, P. and P. Singer. 1993. The great ape project—and beyond. In P. Cavalieri and P. Singer, eds., *The great ape project: Equality beyond humanity*, pp. 304–12. New York: St. Martin's Press.

Cavalieri, P., P. Singer, and contributors. 1993. A declaration on great apes. In P. Cavalieri and P. Singer, eds., *The great ape project: Equality beyond humanity*, pp. 4–7. New York: St. Martin's Press.

Carvalho, J. E. 1992. Relatório de contato com um grupo Guajá, maio/junho 1992. Ministério da Justiça, Fundação Nacional do Índio.

Castro, R. and P. Soini. 1978. Field studies on Saguinus mystax and other callitrichids in Amazonian Peru. In D. G. Kleiman, ed., *The biology and conservation of the callitrichidae*, pp. 73–78. Washington: Smithsonian Institution Press.

Cavalcante, P. B. 1996. *Frutas comestíveis da Amazônia.* Belém: Museu Paraense Emílio Goeldi.

Centro Ecumênico de Documentação e Informação/Projeto Estudo sobre Terras Indígenas no Brasil (CEDI/PETI). 1990. *Terras indígenas no Brasil.* Rio de Janeiro: Museu Nacional.

Center for Disease Control. 1998. Fatal cercopithecine herpesvirus 1 (B virus) infection following a mucocutaneous exposure and interim recommendations for worker protection. *Morbidity and Mortality Weekly Report* 47:1073–76, 1083.

———. 1997. Nonhuman primate spumavirus infections among persons with occupational exposure—United States, 1996. *Morbidity and Mortality Weekly Report* 46:129–31.

———. 1987a. Epidemiologic Notes and Reports B-Virus Infection in Humans—Pensacola, Florida. *Mortality and Morbidity Weekly Report* 36: 289–90, 295–96.

———. 1987b. Guidelines for prevention of *Herpesvirus simiae* (B virus) infection in monkey handlers. *Morbidity and Mortality Weekly Report* 36: 680–82, 687–89.

Chagnon, N. A. 1997. *Yanomamö.* 5th ed. New York: Holt, Rinehart, and Winston.

Ciochon, R. L. and A. B. Chiarelli. 1980. Paleobiogeographic perspectives on the origin of platyrrhini. In R. L. Ciochon and A. B. Chiarelli, eds., *Evolutionary biology of the New World monkeys and continental drift*, pp. 459–93. New York: Plenum Press.

Chiarella, A. G. 1993. Home range of the brown howler monkey, *Alouatta fusca*, in a forest fragment of southeastern Brazil. *Folia Primatologia* 60:173–75.

Chivers, D. J. 1977. The lesser apes. In His Serene Highness Prince Ranier III of Monaco and G. H. Bourne, eds., *Primate conservation*, pp. 539–98. New York: Academic Press.

Chodorow, N. 1974. Family structure and feminine personality. In M. Z. Rosaldo and L. Lamphere, eds., *Women, culture, and society*, pp. 43–67. Stanford: Stanford University Press.

Clarke, M. R. and K. E. Glander. 1984. Female reproductive success in a group of free-ranging howling monkeys (*Alouatta palliata*) in Costa Rica. In M. F. Small, ed., *Female primates: Studies by women primatologists*, pp. 111–26. New York: Allan R. Liss.

Clarke, M. R. and E. L. Zucker. 1994. Survey of the howling monkey population at La Pacifica: A seven year follow-up. *International Journal of Primatology* 15:29–41.

Clarke, M. R., K. E. Glander, and E. L. Zucker. 1998. Infant-nonmother interactions of free-ranging mantled howlers (*Alouatta palliata*) in Costa Rica. *International Journal of Primatology* 19:451–71.

Clarke, M. R., E. L. Zucker, and K. E. Glander. 1994. Group takeover by a natal male howler monkey (*Alouatta palliata*) and the associated disappearance and injuries of immatures. *Primates* 35:435–42.

Clarkson, T. B. and S. A. Klumpp. 1990. The contribution of nonhuman primates to understanding coronary artery atherosclerosis in humans. *Institute for Laboratory Animal Research* 32:4–8.

Clastres, P. 1972. Guayakí cannibalism. In P. J. Lyon, ed., *Native South Americans: Ethnology of the least known continent*, pp. 308–21. Boston: Little, Brown and Company.

——. 1998. *Chronicle of the Guayaki Indians*. New York: Zone Books.

Cleary, David. 1998. "Lost altogether to the civilised world": Race and the Cabanagem in northern Brazil, 1750–1850. *Comparative Studies in Society and History* 40:109–35.

Clutton-Brock, J. 1989. *A natural history of domesticated animals*. Austin: University of Texas Press.

Coelho, P. O. 1994a. Relatório de atividades do PIN Awá referente ao mêses de abril, maio e junho de 1994. Ministério da Justiça, Fundação Nacional do Índio.

——. 1994b. Relatório de atividades do PIN Awá referente ao mês de julho, agosto e setembro 1994. Ministério da Justiça, Fundação Nacional do Índio.

——. 1994c. Relatório de atividades do PIN Awá referente ao mês de outubro, novembro e dezembro de 94. Ministério da Justiça, Fundação Nacional do Índio.

Coimbra-Filho, A. F., I. G. Câmara, and A. B. Rylands. 1995. On the geographic distribution of the red-handed howling monkey, *Alouatta belzebul*, in north-east Brazil. *Neotropical Primates* 3:176–78.

Coimbra-Filho, A. F. and R. A. Mittermeier. 1977. Conservation of the Brazilian lion tamarins (*Leontopithecus rosalia*). In His Serene Highness Prince Ranier III of Monaco and G. H. Bourne, eds., *Primate conservation*, pp. 59–94. New York: Academic Press.

Cole, T. M. III. 1992. Postcranial heterochrony of the masticatory apparatus in *Cebus apella* and *Cebus albifrons*. *Journal of Human Evolution* 23:253–82.

Colinvaux, P. A. and M. B. Bush. 1991. The rain-forest ecosystem as a resource for hunting and gathering. *American Anthropologist* 93:153–160.

Collins, W. E. 1994. The owl monkey as a model for malaria. In J. F. Baer, R. E. Weller, and I. Kakoma, eds., *Aotus: The owl monkey*, pp. 217–44. San Diego: Academic Press.

Comaroff, J. and J. L. Comaroff. 1990. Goodly beasts, beastly goods: cattle and commodities in a South African context. *American Ethnologist* 17:196–216.

Conklin, B. A. 1995. "Thus are our bodies, thus was our custom": Mortuary cannibalism in an Amazonian society. *American Ethnologist* 22:75–101.

——. 1997. Consuming images: Representations of cannibalism on the Amazonian frontier. *Anthropological Quarterly* 70:68–78.

——. 2001. *Consuming Grief: Compassionate cannibalism in an Amazonian society.* Austin: University of Texas Press.

Cooke, H. B. S. 1963. Pleistocene mammal faunas of Africa, with particular reference to southern Africa. In F. C. Howell and F. Bourlière, eds., *African ecology and human evolution*, pp. 65–116. New York: Wenner-Gren Foundation for Anthropological Research.

Cormier, L. A. 1999. Ritualized remembering and genealogical amnesia. *Southern Anthropologist* 26:31–41.

——. 2002a. Monkey as food, monkey as child: Guajá symbolic cannibalism. In A. Fuentes and L. Wolfe, eds., *Primates face to face: The conservation implications of human-nonhuman primate interconnections*, pp. 63–84. Cambridge: Cambridge University Press.

——. 2002b. Monkey ethnobotany: Preserving biocultural diversity in Amazonia. In J. R. Stepp, F. Wyndham, and R. Zarger, eds., *Ethnobiology and biocultural diversity: Proceedings of the seventh international congress of ethnobiology* pp. 313–25. Athens: University of Georgia Press.

——. In press. Decolonizing history: ritual transformation of the past by the Guajá Indians of eastern Amazonia. In N. L. Whitehead, ed., *History and historicities: New perspectives in Amazonia*. Lincoln: University of Nebraska Press.

Correia de Alencar, E. 1991. Funai reencontra os Amanayé. In *Provos indigenas no Brasil, 1987/88/89/90*, p. 346. São Paulo: Centro Ecumênico de Documentação e Informação.

Cristina dos Santos, M., M. Martins, A. L. Luiz Boechat, R. Pereira de Sá-Neto, and M. Ermelinda de Oliveira. 1995. *Serpentes de interesse médico da Amazônia: Biologia, venenos, e tratamento de acidentes.* Manaus: Universidade de Amazonas.

Crocker, J. C. 1985. *Vital souls: Bororo cosmology, natural symbolism, and shamanism.* Tucson: University of Tucson Press.

Crockett, C. M. 1998. Conservation biology for the Genus *Alouatta*. *International Journal of Primatology* 19:549–78.

Crockett, C. M. and J. F. Eisenberg. 1986. Howlers: variations in group size and demography. In B. B. Smuts, R. M. Seyfarth, R. W. Wrangham, and T. T. Struhsaker, eds., *Primate societies*, pp. 54–68. Chicago: University of Chicago Press.

Cruz, E. 1963. *História do Pará.* Vol. 1. Belém: Universidade Federal do Pará.

Cunha, P. 1987. Análise fonêmica preliminar da língua Guajá. Master's thesis, Universidade Estadual de Campinas.

Cyranoski, D. 2000. Row over fate of endangered monkeys. *Nature* 408:280.

Damasceno da Silva, J. A. 1989. Relatório de atividades de saúde desenvolvida junto a communidade Guajá, referente ao mes de novembro. Ministério da Justiça, Fundação Nacional do Índio.

——. 1990. Relatório de atividades desenvolvida no PINC Awá, referente ao mes de Março. Ministério da Justiça, Fundação Nacional do Índio.

——. 1996a. Relatório de atividades, Posto Indígena Awá de janeiro a março 1996. Ministério da Justiça, Fundação Nacional do Índio.

——. 1996b. Relatório de atividades, Posto Indígena Awá de abril a julho 1996. Ministério da Justiça, Fundação Nacional do Índio.

——. 1996c. Relatório de atividades, Posto Indígena Awá de agosto a outubro 1996. Ministério da Justiça, Fundação Nacional do Índio.

——. 1996d. Relatório de atividades, Posto Indígena Awá, novembro/dezembro 1996. Ministério da Justiça, Fundação Nacional do Índio.

Deag, J. M. 1977. The status of the Barbary macaque *Macaca sylvanus* in captivity and factors influencing its distribution in the wild. In His Serene Highness Prince Ranier III of Monaco and G. H. Bourne, eds., *Primate conservation*, pp. 267–87. New York: Academic Press.

Defoliart, G. 1988. Insects as food in indigenous populations. In D. A. Posey and W. L. Overal, eds., *Ethnobiology: Implications and applications. Proceedings of the first international congress of ethnobiology (Belém, 1988).* Vol. 1, pp. 145–50. Belém: Museu Paraense Emílio Goeldi.

Delson, E. and A. L. Rosenberger. 1980. Phyletic perspectives on platyrrhine origins and anthropoid relationships. In L. Ciochon and A. B. Chiarelli, eds., *Evolutionary biology of the New World monkeys and continental drift*, pp. 445–58. New York: Plenum Press.

Descola, P. 1992. Societies of nature and the nature of society. In Adam Kuper, ed., *Conceptualizing society*, pp. 107–26. New York: Routledge.

——. 1994. Pourquoi les Indiens d'Amazonie n'ont-ils pas domestiqué le pécari? Généalogie des objets et anthropologie de l'objectivation. In B. Latour and P.

Lemonnier, eds., *De la préhistoire aux missiles balistiques: L'Intelligence sociale des techniques*, pp. 329–44. Paris: Éditions La Découverte.

——. 1996. Constructing natures: Symbolic ecology and social practice. In P. Descola and G. Pálsson, eds., *Nature and society: anthropological perspectives*, pp. 82–102. New York: Routledge.

——. 1998. Estrutura ou sentimento: a relação com o animal na Amazônia. *Mana* 4:23–45.

de Waal, F. 1989. *Peacemaking among primates*. Cambridge: Harvard University Press.

Diamond, J. 1993. *The third chimpanzee: The evolution and future of the human animal*. New York: Harper Perennial.

Dillehay, T. D. 1989. *Monte Verde: A late Pleistocene settlement in Chile*. Vol. 1. Washington: Smithsonian Institution Press.

——. 1997. *Monte Verde: A late Pleistocene settlement in Chile. The Archaeological Context and Interpretation*. Vol. 2. Washington: Smithsonian Institution Press.

Dittus, W. P. J. 1977. The socioecological basis for the conservation of the Toque monkey (*Macaca sinica*) of Sri Lanka (Ceylon). In His Serene Highness Prince Ranier III of Monaco and G. H. Bourne, eds., *Primate conservation*, pp. 237–65. New York: Academic Press.

Dobroruka. L. J. 1972. Social communication in the brown capuchin, *Cebus apella*. *International Zoo Yearbook* 12:43–45.

Dodt, G. 1939 [1873]. *Descripção dos Rios Parnahyba e Gurupy*. Coleção Brasiliana. Vol. 138. São Paulo: Companhia Editora Nacional.

Dole, G. 1972. Endocannibalism among the Amahuaca Indians. In P. J. Lyon, ed., *Native South Americans: Ethnology of the least known continent*, pp. 302–8. Boston: Little, Brown and Company.

Dufour, D. 1987. Insects as food: A case study from the northwest Amazon. *American Anthropologist* 89:383–97.

Du Mond, F. V. and T. C. Hutchinson. 1967. Squirrel monkey reproduction: The "fatted" male phenomenon and seasonal spermatogenesis. *Science* 158:1067–70.

Dumont, L. 1953. The Dravidian kinship terminology as an expression of marriage. *Man* 54:34–39.

Dunbar, R. I. M. 1977. The gelada baboon: status and conservation. In His Serene Highness Prince Ranier III of Monaco and G. H. Bourne, eds., *Primate conservation*, pp. 363–83. New York: Academic Press.

Durham, N. 1975. Some ecological, distributional, and group behavioral features of Atelinae in southern Peru: with comments on interspecific relations. In R. H. Tuttle, ed., *Socioecology and psychology of primates*, pp. 87–101. The Hague: Mouton.

Durkheim, E. 1995 [1912]. *The elementary forms of the religious life*. Trans. K. E. Fields. New York: The Free Press.

Eberle, R. and J. Hilliard. 1995. The simian herpesviruses. *Infectious Agents and Disease* 4:55–70.

Eisenberg, J. F. 1978. Comparative ecology and reproduction of New World monkeys. In D. G. Kleiman, ed., *The biology and conservation of the callitrichidae*, pp. 13–22. Washington: Smithsonian Institution Press.

Eisenberg, J. F., M. A. Muchkenhirn, and R. Rudran. 1972. The relation between ecology and social structure in primates. *Science* 176:863–74.

Emmons, L. H. and F. Feer. 1990. *Neotropical rainforest mammals: A field guide*. Chicago and London: University of Chicago Press.

——. 1997. *Neotropical rainforest manuals: A field guide*. 2nd ed. Chicago and London: University of Chicago Press.

Erikson, P. 2000. The social significance of pet-keeping among Amazonian Indians. In P. Poberseck and J. Serpell, eds., *Companion Animals and Us*, pp. 7–26. Cambridge: Cambridge University Press.

——. In press. Several fathers in one's cap . . . . Polyandrous conception among the Panoan Matis (Amazonas, Brazil). In S. Beckerman and P. Valentine, eds., *Cultures of multiple fathers: The theory and practice of partible paternity in South America*. Gainesville: University Press of Florida.

Ernst, T. M. 1999. Onobasulu cannibalism and the moral agents of misfortune. In L. R. Goldman, ed., *The anthropology of cannibalism*, pp. 143–59. Westport, Conn.: Bergin and Garvey.

Eudey, A. A. 2002. The primatologist as minority advocate. In A. Fuentes and L. D. Wolfe, eds., *Primates face to face: Conservation implications of human-nonhuman primate interconnections*, pp. 277–87. Cambridge: Cambridge University Press.

Evans-Pritchard, E. E. 1940. *The Nuer: A description of the modes of livelihood and political institutions of a Nilotic people*. Oxford: Clarendon Press.

Fausto, C. 1997. A dialética da predação e familiarização entre os Parakanã da Amazônia Oriental: por uma teoria da guerra ameríndia. Ph. D. dissertation. Universidade Federal do Rio de Janeiro. Museu Nacional.

——. 1999. Of enemies and pets: warfare and shamanism in Amazonia. *American Ethnologist* 26:933–56.

Feldman, D. A. 1990. Assessing viral, parasitic, and sociocultural cofactors affecting HIV-1 transmission in Rwanda. In D. A. Feldman, ed., *Culture and AIDS*, pp. 45–54. New York: Praeger.

Feng, H. Y. 1936. Teknonymy as a formative factor in the Chinese kinship system. *American Anthropologist* 38:59–66.

Gao, F., E. Bailes, D. L. Robertson, Y. Chen, C. M. Rodenburg, S. F. Michael, L. B. Cummins, L. O. Arthur, M. Peeters, G. M. Shaw, P. M. Sharp, and B. H. Hahn. 1999. Origin of HIV-1 in the chimpanzee *Pan troglodytes troglodytes*. *Nature* 397:436–41.

Ferrari, S. F. and A. P. de Souza Jr. 1994. More untufted capuchins in southeastern Amazonia? *Neotropical Primates* 2:9–10.

Ferrari, S. F. and M. A. Lopes. 1996. Primate populations in eastern Amazonia. In M. A. Norconk, A. L. Rosenberger, and P. A. Garber, eds., *Adaptive radiations of Neotropical primates*, pp. 53–67. New York and London: Plenum Press.

Fiennes, R. 1967. *Zoonoses of primates: The epidemiology and ecology of simian diseases in relation to man*. Ithaca: Cornell University Press.

Fleagle, J. G. 1988. *Primate adaptation and evolution*. San Diego: Academic Press.

——. 1999. *Primate adaptation and evolution*. 2nd ed. San Diego: Academic Press.

Fleagle, J. G. and R. F. Kay. 1997. Platyrrhines, catarrhines, and the fossil record. In W. G. Kinzey, ed., *New World primates: Ecology, evolution, and behavior*, pp. 2–23. Chicago: Aldine.

Fleagle, J. G. and R. A. Mittermeier. 1980. Locomotor behavior, body size, and comparative ecology of seven Surinam monkeys. *American Journal of Physical Anthropology* 52:301–14.

Flynn, J. J. and A. R. Wyss. 1998. Recent advances in South American mammalian paleontology. *Trends in ecology and evolution* 13: 449–469.

Fonesca de Castro, A. 1996. Manuscritos sobre a Amazônia Colonial: Repertorio referente a mâo-de-obra indígena do fundo secretaria do governo (Colônia e Império). *Anais do Arquivo Público do Pará* 2 (1): 9–121.

Fooden, J. 1963. A revision of the woolly monkeys (genus Lagothrix). *Journal of Mammalogy* 44:213–47.

Ford, S. 1994. The taxonomy and distribution of the owl monkey. In J. F. Gaer, R. E. Weller, and I. Kakoma, eds., *Aotus: The owl monkey*, pp. 1–57. San Diego: Academic Press.

Ford, S. M. 1986. Subfossil platyrrhine tibia (Primates: Callitrichidae) from Hispaniola: A possible further example of island gigantism. *American Journal of Physical Anthropology* 70:47–62.

——. 1990. Platyrrhine evolution in the West Indies. *Journal of Human Evolution* 19:237–54.

Forline, L. C. 1997. The persistence and cultural transformation of the Guajá Indians: foragers of Maranhão State, Brazil. Ph.D. diss., University of Florida. Ann Arbor: UMI Dissertation Services.

Forthman-Quick, D. L. 1984. Activity budgets and the consumption of human food in two troops of baboons, *Papio anubis*, at Gilgil, Kenya. In J. G. Else and P. C. Lee, eds., *Primate ecology and conservation*, Cambridge: Cambridge University Press.

Fouts, R. S., D. H. Fouts, and G. Waters. 2002. The ethics and efficacy of biomedical research in chimpanzees with special regard to HIV research. In A. Fuentes and L. D. Wolfe, eds., *Primates face to face: Conservation implications of human-nonhuman primate interconnections*, pp. 45–60. Cambridge: Cambridge University Press.

Fragaszy, D. M. 1978. Contrasts in feeding behavior in squirrel and titi monkeys. In D .J. Chivers and J. Herbert, eds., *Recent advances in primatology. Vol. 1: Behavior*, pp. 363–68. New York: Academic Press.

Fragaszy, D. M. and E. Visalberghi 1989. Social influences on the acquisition and use of tools in tufted capuchin monkeys (*Cebus apella*). *Journal of Comparative Psychology* 103:159–70.

Freese, C. and J. R. Oppenheimer. 1981. The capuchin monkey, genus *Cebus*. In A. F. Coimbra-Filho and R. A. Mittermeier, eds., *Ecology and behavior of Neotropical primates. Vol. 1*, pp. 331–90. Rio de Janeiro: Academia Brasileira de Ciências.

Freese, C. H., M. A. Freese, and N. R. Castro. 1978. The status of callitrichids in Peru. In D. G. Kleiman, ed., *The biology and conservation of the callitrichidae*, pp. 121–30. Washington: Smithsonian Institution Press.

French, J. A., D. H. Abbott, and C. T. Snowdon. 1984. The effect of social environment on estrogen secretion, scent-marking, and sociosexual behavior in tamarins (*Saguinus oedipus*). *American Journal of Primatology* 6:155–67.

Fuentes, A. 2002. Monkeys, humans and politics in the Mentawai Islands: No simple solutions in a complex world. In A. Fuentes and L. D. Wolfe, eds., *Primates face to face: Conservation implications of human-nonhuman primate interconnections*, pp. 187–207. Cambridge: Cambridge University Press.

Fuentes A. and L. D. Wolfe. 2002. Introduction. In *Primates face to face: Conservation implications of human-nonhuman primate interconnections*, pp. 1–4. Cambridge: Cambridge University Press.

Galland, G. G. 2000. Role of the squirrel monkey in parasitic disease research. *Institute of Laboratory Animal Research* 41:37–43.

Gardner, D. 1999. Anthropophagy, myth, and the subtle ways of ethnocentrism. In L. R. Goldman, ed., *The anthropology of cannibalism*. pp. 27–49. Westport, Conn.: Bergin and Garvey.

Gery, J. 1977. *Characoids of the world*. Neptune City, N.J.: T. F. H. Publications.

Geertz, C. 1973. Deep play: Notes on the Balinese cockfight. In *The interpretation of cultures*, pp. 412–53. New York: Basic Books.

Geertz, H. and C. Geertz. 1964. Teknonymy in Bali: Parenthood, age-grading and genealogical amnesia. *Journal of the Royal Anthropological Institute* 94:94–108.

Gilmore, R. M. 1950. Fauna and ethnozoology of South America. In J. H. Steward, ed., *Handbook of South American Indians*. Vol. 6: *Physical anthropology, linguistics, and cultural geography of South American Indians*. Bureau of American Ethnology Bulletin 143:345–464. Washington: Smithsonian Institution Press.

Gingerich, P. D. 1980. Eocene Adapidae, paleobiogeography, and the origin of South American platyrrhini. In R. L. Ciochon and A. B. Chiarelli, eds., *Evolutionary Biology of the New World monkeys and continental drift*, pp. 123–38. New York: Plenum Press.

Godelier, M., T. R. Trautmann, F. E. Tjon Sie Fat, eds. 1998. *Transformations of kinship*. Washington: Smithsonian Institution Press.

Goldman, I. 1948. Tribes of the Uaupes-Caqueta region. In J. H. Steward, ed., *Handbook of South American Indians*. Vol. 3: *The Tropical Forest Tribes. Bureau of American Ethnology Bulletin* 143:793–98. Washington: Smithsonian Institution Press.

Goldman, L. R. 1999. From pot to polemic: uses and abuses of cannibalism. In L. R. Goldman, ed., *The anthropology of cannibalism*, pp. 1–26. Westport, Conn.: Bergin and Garvey.

Gomes, M. P. 1985a. Relatório antropológico sobre a Área Indígena Guajá (Awá-Gurupi). Setembro. Ministério da Justiça, Fundação Nacional do Índio.

——. 1985b. Relatório sobre os índios Guajá próximos à Ferrovia Carajás km 400. Ministério da Justiça, Fundação Nacional do Índio.

——. 1988. O povo Guajá e as condiçoes reais para à sua sobrevivencia: reflexões e propostas. Ministério da Justiça, Fundação Nacional do Índio.

——. 1991. O povo Guajá e as condições reais para a sua sobrevivência. In *Povos indígenas no Brasil 1987/88/89/90*, pp. 354–60. São Paulo: Centro Ecumênico de Documentação e Informação.

——. 1996. Os índios Guajá: demografia, terras, perspectivas de futuro. Relatório de pesquisas realizadas em fevereiro de 1996. Unpublished manuscript.

Good, A. 1980. Elder sister's daughter marriage in south Asia. *Journal of Anthropological Research* 36:474–500.

Goodall, A. G. and C. P. Groves. 1977. The conservation of eastern gorillas. In His Serene Highness Prince Ranier III of Monaco and G. H. Bourne, eds., *Primate conservation*, pp. 599–637. New York: Academic Press.

Goodall, J. V. 1971. *In the shadow of man*. London: Collins.

Goodman, S. 2001. Europe brings experiments on chimpanzees to an end. *Nature* 411:123.

Goulding, M. 1980. *The fishes and the flooded forest: Explorations in Amazonian natural history*. Berkeley: University of California Press.

Gragson, T. L. 1992. Fishing the waters of Amazonia: Native subsistence economies in a tropical rain forest. *American Anthropologist* 94:428–40.

Graham, L. R. 1995. *Performing dreams: Discourses of immortality among the Xavante of central Brazil*. Austin: University of Texas Press.

Green, K. M. 1976. The nonhuman primate trade in Colombia. In R. W. Thorington Jr. and P. B. Heltne, eds., *Neotropical primates: Field studies and conservation*, pp. 85–95. Washington: National Academy of Sciences.

Gregor, T. A. 1974. Publicity, privacy, and marriage. *Ethnology* 13:333–49.

——. 1977. *Mehinaku: The drama of daily life in a Brazilian Indian village*. Chicago: University of Chicago Press.

——. 1985. *Anxious pleasures: The sexual lives of an Amazonian people*. Chicago: University of Chicago Press.

Grenand, P. and F. Grenand. 1981. La médecine traditionnelle des Wayãpi, améridiens de Guyane. *Cahiers Orstom* 18:561–67.

Gribbin, J. 1984. *In search of Schrödinger's cat: Quantum physics and reality*. New York: Bantam Books.

Gross, D. R. 1975. Protein capture and cultural development in the Amazon basin. *American Anthropologist* 77:526–549.

Gross, D. R., G. Eiten, N. M. Flowers, F. M. Leoi, M. L. Ritter, and D. W. Werner. 1979. Ecology and acculturation among native peoples of Central Brazil. *Science* 206:1043–50.

Groves, C. P. 1981. Comments to P. Shipman, W. Bosler, and K. L. Davis, "Butchering of giant baboons at an Acheulian site." *Current Anthropolgy* 22:265.

Halbmayer, E. 2001. The dead, their souls, and the living among the Carib-speaking Indians: Comparing relations of alterity. Paper presented at the American Anthropological Association invited session: Present and Future Research Directions Among Indigenous Groups in the Guianas and its Neighbors: An Evaluation. Washington, November 2001.

Happel, R. E., J. F. Noss, and C. W. Marsh. 1987. Distribution, abundance, and endangerment of primates. In C. W. Marsh and R. A. Mittermeier, eds., *Primate conservation in the tropical rain forest: Monographs in primatology*. Vol. 9, pp. 63–82. New York: Alan R. Liss.

Haraway, D. J. 1989. *Primate visions: Gender, race, and nature in the world of modern science*. New York: Routledge.

Harris, M. 1968. *The Rise of anthropological theory: A history of theories of culture*. New York: HarperCollins.

——. 1974. *Cows, pigs, wars, and witches*. New York: Random House.

Hart, T. B. and J. A. Hart. 1986. The ecological basis of hunter-gatherer subsistence in African rain forests: The Mbuti of eastern Zaire. *Human Ecology* 14:29–56.

Hartwig, W. C. 1995. A giant New World monkey from the Pleistocene of Brazil. *Journal of Human Evolution* 28:189–95.

Hartwig, W. C. and C. Cartelle. 1996. A complete skeleton of the giant South American primate *Protopithecus*. *Nature* 381:307–11.

Headland, T. N. 1987. The wild yam question: How well could independent hunter-gatherers live in a tropical rain forest ecosystem? *Human Ecology* 15:463–491.

Heltne, P. G. and R. W. Thorington Jr. 1976. Problems and potentials for primate biology and conservation in the New World. In R. W. Thorington Jr. and P. B. Heltne. eds., *Neotropical primates: Field studies and conservation*, pp. 110–24. Washington: National Academy of Sciences.

Hernández-Camacho, J. and R. W. Cooper. 1976. The nonhuman primates of Colombia. In R. W. Thorington Jr. and P. B. Heltne, eds., *Neotropical primates: Field studies and conservation*, pp. 35–69. Washington: National Academy of Sciences.

Herndon, W. L. 2000. *Exploration of the valley of the Amazon, 1851–1852*. Ed. G. Kinder. New York: Grove Press.

Hershkovitz, P. 1972. Notes on New World monkeys. *International Zoo Yearbook* 12:3–12.

——. 1977. *Saguinus midas* group, Midas tamarins, *Saguinus midas* Linnaeus. In *Living New World monkeys (Platyrrhini)*. Vol. 1, pp. 706–31. Chicago: University of Chicago Press.

——. 1983. Two new species of night monkeys, genus *Aotus* (Cebidae, Platyrrhini): a preliminary report on *Aotus* taxonomy. *American Journal of Primatology* 4:209–43.

——. 1984. Taxonomy of squirrel monkeys genus *Saimiri* (Cebidae, Platyrrhini): a preliminary report with description of a hitherto unnamed form. *American Journal of Primatology* 7:155–210.

Heymann, E. W. 1998. Giant fossil New World primates: arboreal or terrestrial? *Journal of Human Evolution* 34:99–101.

Hill, C. 2000. Conflict of interest between people and baboons: crop raiding in Uganda. *International Journal of Primatology* 21: 299–315.

Hill, K. and K. Hawkes. 1983. Neotropical hunting among the Aché, of Eastern Paraguay. In R. B. Hames and W. T. Vickers, eds., *Adaptive responses of native Amazonians*, pp. 139–88. New York: Academic Press.

Hill, K., K. Hawkes, M. Hurtado, and H. Kaplan. 1984. Seasonal variance in the diet of Aché, hunter-gatherers in eastern Paraguay. *Human Ecology* 12:101–35.

Hladik, C. M. 1978. Adaptive strategies of primates in relation to leaf-eating. In G. G. Montgomery, ed., *The ecology of arboreal folivores*, pp. 373–95. Washington: Smithsonian Institution Press.

Hoffstetter, R. 1980. Origin and deployment of New World monkeys emphasizing the southern continents route. In R. L. Ciochon and A. B. Chiarelli, eds., *Evolutionary biology of the New World monkeys and continental drift*, pp. 102–21. New York: Plenum Press.

Holmberg, A. R. 1985 [1950]. *Nomads of the long bow: The Siriono of eastern Bolivia*. Prospect Heights, Ill.: Waveland Press.

Hornborg, A. 1998. Serial redundancy in Amazonian social structure: Is there a method for poststructuralist comparison? In M. Godelier, T. R. Trautmann, and F. E. Tjon Sie Fat, eds., *Transformations of kinship*, pp. 168–86. Washington: Smithsonian Institution Press.

Horwich, R. H. 1998. Effective solutions for howler conservation. *International Journal of Primatology* 19:579–98.

Houle, A. The origin of the platyrrhines: An evaluation of the Antarctic scenario and the floating island model. *American Journal of Physical Anthropology* 109:541–59.

Howell, F. C., L. S. Fichter, and G. Eck. 1969. Vertebrate assemblages from the Usno Formation, White Sands and Brown Sands localities, Lower Omo Basin. *Quaternaria* 11:65–88.

Howell, S. 1996. Nature in culture or culture in nature? Chewong ideas of "humans" and other species. In P. Descola and G. Pálsson, eds., *Nature and society: Anthropological perspectives*, pp. 127–44. New York: Routledge.

Hoyer, B. H., N. W. van de Velde, M. Goodman, and R. B. Roberts. 1972. Estimation of hominid evolution by DNA sequence homology. *Journal of Human Evolution* 1:645–49.

Huffman, M. A. 1997. Current evidence for self-medication in primates: A multidisciplinary perspective. *Yearbook of Physical Anthropology* 40: 171–200.

Hugh-Jones, S. 1979. *The palm and the pleiades: Initiation and cosmology in northwest Amazonia*. Cambridge: Cambridge University Press.

Hunter, J., R. D. Martin, A. F. Dixon, and B. C. Rudder. 1979. Gestation and interbirth intervals in the owl monkey (*Aotus trivirgatus griseimembra*). *Folia Primatologia* 31:165–75.

Hurtado, A. M., K. Hawkes, K. Hill, and H. Kaplan. 1985. Female subsistence strategies among Aché hunter-gatherers of eastern Paraguay. *Human Ecology* 13:1–28.

Hutchins, M. 1999. Conservation organizations, zoological parks, animal welfare, advocates, and medical researchers call for immediate action to address the commercial bushmeat crisis in tropical African countries. *American Society of Primatologists Bulletin* 23:7.

Hutterer, K. L. 1983. The natural and cultural history of Southeast Asian agriculture: Ecological and evolutionary considerations. *Anthropos* 78:169–212.

Ingold, T. 1980. *Hunters, pastoralists, and ranchers: Reindeer economies and their trans-formations.* Cambridge: Cambridge University Press.

——. 1999. On the social relations of the hunter-gatherer band. In R. B. Lee and R. Daly, eds., *The Cambridge encyclopedia of hunters and gatherers,* pp. 399–410. Cambridge: Cambridge University Press.

Isaac, G. L . 1977. *Olorgesaile: archaeological studies of a middle Pleistocene lake basin in Kenya.* Chicago: University of Chicago Press.

Izawa, K. 1975. Foods and feeding behavior of monkeys in the upper Amazon basin. *Primates* 16:295–316.

——. 1976. Group sizes and compositions of monkeys in the upper Amazon basin. *Primates* 17:367–99.

Izawa, K. and A. Mizuno. 1977. Palm-fruit cracking behavior of wild black-capped capuchin *(Cebus apella). Primates* 18:773–92.

Izawa, K. and G. Bejarano. 1981. Distribution ranges and patterns of nonhuman primates in western Pando, Bolivia. *Kyoto University Overseas Research Reports of New World Monkeys* 2:13–22.

Jackson, J. E. 1983. *The fish people: Linguistic exogamy and Tukanoan identity in northwest Amazonia.* Cambridge: Cambridge University Press.

Janson, C. H. and S. Boinski. 1992. Morphological and behavioral adaptations for foraging in generalist primates: The case of the cebines. *American Journal of Physical Anthropology* 88:483–98.

Jensen, C. 1999. Tupí-Guaraní. In R. M. Dixon and A. Y. Aikhenvald, eds., *The Amazonian languages,* pp. 125–63. Cambridge: Cambridge University Press.

Johns, A. D. and J. M. Ayres. 1987. Southern bearded sakis beyond the brink. *Oryx* 21: 164–67.

Johnson, A. 1975. Time allocation in a Machiguenga community. *Ethnology* 14:301–10.

Johnson, K. M. 1993. Emerging viruses in context: An overview of viral hemorrhagic fevers. In S. S. Morse, ed., *Emerging viruses,* pp. 46–57. New York: Oxford University Press.

Jucá Filho, R. 1987. Relatório de 20 Novembre, 1987. Ministério da Justiça, Fundação Nacional do Índio.

Julliot, C. 1996. Seed dispersal by red howling monkeys *(Alouatta seniculus)* in the tropical rainforest of French Guiana. *International Journal of Primatology* 17:239–58.

Kalter, S. S. 1977. The baboon. In His Serene Highness Prince Ranier III of Monaco and G. H. Bourne, eds., *Primate conservation,* pp. 385–418. New York: Academic Press,

Kantner, J. 1999. Anasazi mutilation and cannibalism in the American Southwest. In L. R. Goldman, ed., *The anthropology of cannibalism,* pp. 75–104. Westport, Conn.: Bergin and Garvey.

Kaplan, H. and K. Kopischke. 1992. Resource use, traditional technology, and change among native peoples of lowland South America. In K. Redford and C. Padoch, eds., *Conservation of Neotropical forests: Working from traditional resource use,* pp. 83–107. Columbia University Press: New York.

Kay, R. F. 1990. The phyletic relationships of extant and fossil Pitheciinae (Platyrrhine, Anthropoidea). *Journal of Human Evolution* 19:175–208.

Kay, R. F., D. Johnson, and D. J. Meldrum. 1998. A New Pitheciin primate from the middle Miocene of Argentina. *American Journal of Primatology* 45:317–36.

——. 1999. Corrigendum. *American Journal of Primatology* 47:347.

Kay, R. F. and R. H. Madden. 1997. Mammals and rainfall: Paleoecology of the middle Miocene at La Venta (Colombia, South America). *Journal of Human Evolution* 32:161–99.

Kensinger, K. M., P. Rabineau, H. Tanner, S. G. Ferguson, and A. Dawson. 1975. *The Cashinahua of Eastern Peru.* Vol. 1 of *Studies in Anthropology and Material Culture*, J. P. Dwyer, ed. Providence: The Haffenferrer Museum of Anthropology, Brown University.

Kessler, P. 1995. Preliminary field study of the red-handed tamarin, *Saguinus midas*, in French Guiana. *Neotropical Primates* 3:184–85.

Khabbaz, R. F., W. Heneine, J. R. George, B. Parekh, T. Rowe, T. Woods, W. M. Switzer, H. M. McClure, M. Murphey-Corb, and T. M. Folks. 1994. Infection of a laboratory worker with simian immunodeficiency virus. *New England Journal of Medicine* 330:172–77.

King, N. W. 1994. The owl monkey in oncogenic virus research. In J. F. Baer, R. E. Weller, and I. Kakoma, eds., *Aotus: The owl monkey*, pp. 245–61. San Diego: Academic Press.

Kinzey, W. G. 1982. Distribution of primates in forest refuges. In G. T. Prance, ed., *Biological diversification in the tropics*, pp. 455–82. New York: Columbia University Press.

——. 1992. Dietary and dental adaptation in the Pithecine. *American Journal of Physical Anthropology* 88:499–514.

——. 1997. Cebus. In W. G. Kinzey, ed., *New World primates: Ecology, evolution, and behavior*, pp. 248–257. Chicago: Aldine.

Klein, L. L. and D. J. Klein. 1976. Neotropical primates: Aspects of habitat usage, population density, and regional distribution in La Macerna, Colombia. In R. W. Thorington Jr. and P. B. Heltne, eds., *Neotropical primates: Field studies and conservation*, pp. 70–79. Washington: National Academy of Sciences.

Koenig, R. 1999. European researchers grapple with animal rights. *Science* 284:1604–6.

Kracke, W. H. 1978. *Force and persuasion: Leadership in an Amazonian society.* Chicago: University of Chicago Press.

——. 1999. A language of dreaming: Dreams of an Amazonian insomniac. *International Journal of Psychoanalysis* 80: 257–71.

Kricher, J. C. 1989. *A Neotropical companion: An introduction to the animals, plants, and ecosystems of the New World Tropics.* Princeton: Princeton University Press.

Lamphere, L. 1993. The domestic sphere of women and the public world of men: The strengths and limitations of an anthropological dichotomy. In C. B. Brettell and C. F. Sargent, eds., *Gender in cross-cultural perspective*, pp. 67–77. Englewood Cliffs, N.J.: Prentice Hall.

Lange, A. 1914. *The lower Amazon: A narrative of explorations in the little known regions of the state of Pará, on the lower Amazon.* New York: G. P. Putnam's Sons.

Laraia, R. 1971. A estrutura do parentesco Tupí. In S. C Gudschinsky ed., *Estudos sobre línguas e culturas indígenas,* pp. 174–212. Brasília: Summer Institute of Linguistics.

Lathrap, D. 1968. The hunting economies of the tropical forest zone of South America: An attempt at historical perspective. In R. Lee and I. Devore, eds., *Man the hunter,* pp. 23–29. Chicago: Aldine.

Lavocat, R. 1980. The implications of rodent paleontology and biogeography to the geographical sources and origin of platyrrhine primates. In R. L. Ciochon and A. B. Chiarelli, eds., *Evolutionary biology of the New World monkeys and continental drift,* pp. 93–102. New York: Plenum Press.

Leacock, E. 1978. Women's status in egalitarian society: Implications for social evolution. *Current Anthropology* 19:247–75.

Leakey, L. S. B. 1965. Mammalian fauna other than Bovidae. In *A Preliminary Report on the Geology and Fauna.* Vol. 1 of *Olduvai Gorge 1951–1961,* pp. 12–36. Cambridge: Cambridge University Press.

Lee, R. B. 1968. What hunters do for a living, or, how to make out on scarce resources. In R. Lee and I. Devore, eds., *Man the hunter,* pp. 30–48. Chicago: Aldine.

Lee, R. B. and R. Daly. 1999. Introduction: foragers and others. In R. B. Lee and R. Daly, eds., *The Cambridge encyclopedia of hunters and gatherers,* pp. 1–19. Cambridge: Cambridge University Press.

Leopold, A. S. 1959. *The wildlife of Mexico.* Berkeley: University of California Press.

Lévi-Strauss, C. 1969 [1949]. *The elementary structures of kinship.* Trans. J. H. Bell, J. R. von Sturmer, and R. Needham. Boston: Beacon Press.

Lippold, L. K. 1977. The douc languar: A time for conservation. In His Serene Highness Prince Ranier III of Monaco and G. H. Bourne, eds., *Primate conservation,* pp. 513–38. New York: Academic Press.

Lizarralde, M. 1997. Ethnoecology of monkeys among the Barí of Venezuela: Perception, use, and conservation. Paper presented at the American Anthropological Association invited session: What Are We Doing Watching Monkeys? Anthropological Perspectives and the Role of Nonhuman Primate Research. November, 1997.

——. 2002. Ethnoecology of monkeys among the Barí of Venezuela: Perception, use, and conservation. In A. Fuentes and L. Wolfe, eds., *Primates face to face: The conservation implications of human-nonhuman primate interconnections,* pp. 85–100. Cambridge: Cambridge University Press.

Lizot, J. 1977. Population, resources, and warfare among the Yanomami. *Man* 12:497–517.

Lopes, M. A. and S. F. Ferrari. 1996. Preliminary observations on the Ka'apor capuchin *Cebus kaapori* Queiroz 1992 from eastern Brazilian Amazonia. *Biological Conservation* 76: 321–24.

Lopes, M. A. and S. F. Ferrari. 2000. Effects of human colonization on the abundance and diversity of mammals in eastern Brazilian Amazonia. *Conservation Biology* 14: 1658–65.

Lúcio de Azevedo, J. 1930. *Os Jesuítas no Grão Pará: Suas missões a colonização.* Coimbra: Imprensa da Universidade.

Marks, G. and W.K. Beatty. 1976. *Epidemics.* New York: Scribner's

Marks, J. 1992. What's old and new in molecular phylogenetics. *American Journal of Physical Anthropology* 85:207–19.

——. 1993. Scientific misconduct: Where "Just Say No" fails. *American Scientist* 81:380–82.

Marks, J., C. W. Schmid, and V. M. Sarich. 1988. DNA hybridization as a guide to phylogeny: Relations of the Hominoidea. *Journal of Human Evolution* 17:769–86.

Marsh, C. W., A. D. Johns, and J. M. Ayres. 1987. Effects of habitat disturbance on rain forest primates. In C. W. Marsh and R. A. Mittermeier, eds., *Primate conservation in the tropical rain forest,* pp. 83–107. New York: Alan R. Liss.

Martin, P. S. 1984. Prehistoric overkill, the global model. In P. S. Martin and R. G. Klein, eds., *Quaternary extinctions,* pp. 354–403. Tucson: University of Arizona Press.

Masanaru, T., F. Anaya, N. Shigehara, and T. Setoguchi. 2000. New fossil materials of the earliest New World monkey, *Branisella boliviana,* and the problem of platyrrhine origins. *American Journal of Physical Anthropology* 111:263–81.

Mascarenhas, B. M., M. de Fátima Cunha Lima, and W. L. Overal. 1992. *Animais da Amazônia: Guia zoológico do Museu Paraense Emílio Goeldi.* Belém: Museu Paraense Emílio Goeldi.

Masterson, T. J. 1995. Morphological relationships between the Ka'apor capuchin (*Cebus kaapori,* Queiroz, 1992) and other male *Cebus* crania: A preliminary report. *Neotropical Primates* 3:165–69.

Mason, W. A. 1978. Ontogeny of social systems. In D. J. Chivers and J. Herbert, eds., *Behavior,* Vol. 1 of *Recent advances in primatology,* pp. 5–15. New York: Academic Press.

May, P. H., A. B. Anderson, M. J. Balick, and J. M. F. Frazão. 1985. Subsistence benefits from the babassu palm (*Orbignya martiana*). *Economic Botany* 39:113–29.

Maybury-Lewis, D. 1967. *Akwe-Shavante Society.* Oxford: Clarendon Press.

McDonald, D. R. 1977. Food taboos: a primitive environmental protection agency (South America). *Anthropos* 72:734–48.

McGuire, M. T. 1974. *The St. Kitts vervet.* Basel: S. Karger.

MacIlwain, C. 1997. Closure of primate lab angers both researchers and critics. *Nature* 390:321.

McNeill, W. H. 1976. *Plagues and peoples.* Garden City, N.Y.: Anchor Press/ Doubleday.

McPhee, R. D. and M. C. Rivero. 1996. Accelerator mass spectrometry ^{14}C age determination for alleged "Cuban spider monkey" *Ateles* (= *Montaneia*) *anthropomorphus. Journal of Human Evolution* 30:89–94.

Meirelles, J. C., Jr., 1973. Contato com os Guajás do Alto Rio Turiaçu. Ministério da Justiça, Fundação Nacional do Índio.

——. 1985. Relatório do reconhecimento da área da Serra da Desordem. Setembro. Ministério da Justiça, Fundação Nacional do Índio.

Meggers, B. J. 1957. Environment and culture in the Amazon basin: An appraisal of the theory of environmental determinism. In A. Palerm ed., *Studies in human ecology: Social science monographs no. III*, pp. 71 -89. Washington: Pan American Union.

———. 1971. Aboriginal adaptation to the terra firme. In *Amazonia: Man and Culture in a counterfeit paradise*, pp. 39–96. Aldine: Chicago.

———. 1973. Some problems of cultural adaptation in Amazonia, with emphasis on the pre-European Period. In B. Meggers, E. Ayensu, and W. Duckworth, eds., *Tropical forest ecosystems in Africa and South America: A comparative review*, pp. 311–20. Washington: Smithsonian Institution Press.

Métraux, A., and H. Baldus. 1946. The Guayaki. In J. H. Steward, ed., *The marginal tribes: Bureau of American ethnology bulletin 143*, pp. 435–44. Vol. 1 of *Handbook of South American indians*. Washington, D.C.: Smithsonian Institution Press.

Meyer de Schauensee, R. 1964. *The birds of Colombia and the adjacent areas of South and Central America*. Narberth, Penn.: Livingston.

Miller, E. R. and E. L. Simons. 1997. Dentition of *Proteopithecus sylviae*, an archaic anthropoid from the Fayum, Egypt. *Proceedings of the National Academy of Sciences of the United States of America* 94:13,760–13,764.

Milton, K. 1980. *The foraging strategy of howler monkeys: A study in primate economics*. New York: Columbia University Press.

———. 1984. Protein and carbohydrate resources of the Maku Indians of northwestern Amazonia. *American Anthropologist* 86:7–27.

Milton, K. A. and J. L. Nessimian. 1984. Evidence for insectivory in two primate species (*Callicebus torquatus* and *Lagothrix lagothricha lagothricha*) from northwestern Amazonia. *American Journal of Primatology* 6:367–71.

Mittermeier, R. A. 1987a. Effects of hunting on rain forest primates In C. W. Marsh and R. A. Mittermeier, eds., *Primate conservation in the tropical rain forest: Monographs in primatology*. Vol. 9, pp. 109–46 New York: Alan R. Liss.

———. 1987b. Framework for primate conservation in the Neotropical region. In C. W. Marsh and R. A. Mittermeier, eds., *Primate conservation in the tropical rain forest: Monographs in primatology*. Vol. 9, pp. 305–20. New York: Alan R. Liss.

———. 1987c. Primate conservation priorities in the Neotropical region. In C. W. Marsh and R. A. Mittermeier, eds., *Primate conservation in the tropical rain forest: Monographs in primatology*. Vol. 9, pp. 221–40. New York: Alan R. Liss.

Mittermeier, R. A., R. C. Bailey, and A. F. Coimbra-Filho. 1978. Conservation status of the Callitrichidae in Brazilian Amazonia, Surinam, and French Guiana. In D. G. Kleiman, ed., *The Biology and Conservation of the Callitrichidae*, pp. 137–46. Washington: Smithsonian Institution Press.

Mittermeier, R. A. and A. F. Coimbra-Filho. 1977. Primate conservation in Brazilian Amazonia. In His Serene Highness Prince Ranier III of Monaco and G. H. Bourne, eds., *Primate conservation*, pp. 117–66. New York: Academic Press.

———. 1981. Systematics: Species and subspecies. In A. F. Coimbra-Filho and R. A. Mittermeier, eds., *Ecology and behavior of Neotropical primates*. Vol. 1, pp. 29–110. Rio de Janeiro: Academia Brasileira de Ciências.

Mittermeier, R. A., W. Kinzey, and R. Mast. 1989. Neotropical primate conservation. *Journal of Human Evolution* 18:597–610.

Mittermeier, R. A., W. R. Konstant, H. Ginsberg, G. M. van Roosmalen, and E. C. da Silva Jr. 1983. Further evidence of insect consumption in the bearded saki monkey *Chiropotes satanas chiropotes*. *Primates* 24:602–5.

Mittermeier, R. A., and M. G. van Roosmalen. 1981. Preliminary observations on habitat utilization and diet in eight Surinam monkeys. *Folia Primatologia* 36:1–39.

Montgomery, E. I. 1970. *With the Shiriana in Brazil.* Dubuque, Iowa: Kendall/Hunt.

Moran, E. F. 1990. Levels of analysis and analytical level shifting: Examples from Amazonian ecosystem research. In E. F. Moran, ed., *The ecosystem approach in anthropology: From concept to practice*, pp. 279–308. Ann Arbor: University of Michigan Press.

——. 1993. *Through Amazonian eyes.* Iowa City: University of Iowa Press.

Morris, R. and D. Morris. 1966. *Men and apes.* New York: McGraw Hill.

Morse, S. S. 1993. Examining the origin of emerging viruses. In S. S. Morse, ed., *Emerging viruses*, pp. 10–28. New York: Oxford University Press.

Moynihan, M. 1976a. *The New World primates.* Princeton, N. J.: Princeton University Press.

——. 1976b. Notes on the ecology and behavior of the pygmy marmoset (Cebuella pygmaeae) in Amazonian Colombia. In R. W. Thorington Jr. and P. B. Heltne, eds., *Neotropical primates: Field studies and conservation*, pp. 79–84. Washington: National Academy of Sciences.

Murphy, B. 1993. Factors restraining emergence of new influenza viruses. In S. S. Morse, ed., *Emerging viruses*, pp. 234–40. New York: Oxford University Press.

Murphy, F. A. 1996. The public health risk of animal organ and tissue transplantation into humans. *Science* 273:746–47.

Murphy, R. F. 1956. Matrilocality and patrilineality in Mundurucú society. *American Anthropologist* 56:414–34.

——. 1960. *Headhunters heritage: Social and economic change among the Mundurucú Indians.* Berkeley and Los Angeles: University of California Press.

Murphy, R. F. and B. Quain. 1966. The Trumaí Indians of central Brazil. In E. S. Goldfrank, ed., *Monographs of the American ethnological society.* No. 24. Seattle: University of Washington Press.

Myerhoff, B. and J. Ruby. 1982. Introduction. In J. Ruby, ed., *A crack in the mirror: Reflexive perspectives in anthropology*, pp. 1–35. Philadelphia: University of Pennsylvania Press.

Myers, G., K. MacInnes, and L. Myers. 1993. Phylogenetic moments in the AIDS epidemic. In S. S. Morse, ed., *Emerging viruses*, pp. 120–37. New York: Oxford University Press.

Nadis, S. 1999. Threat to U.S. primate researchers. *Nature* 402:7–8.

Nakatsukasa, M., M. Takai, and T. Setoguchi. 1997. Functional morphology of the postcranium and locomotor behavior of *Neosaimiri fieldsi*, a *Saimiri*-like middle Miocence Platyrrhine. *American Journal of Physical Anthropology* 102:515–44.

Napier, J. R. and P. H. Napier. 1985. *The natural history of primates*. Cambridge, Mass.: MIT Press.

Needham, R. 1954. The system of teknonyms and death-names of the Penan. *Southwestern Journal of Anthropology* 10:416–31.

Neville, M. 1976. The population and conservation of howler monkeys in Venezuela and Trinidad. In R. W. Thorington Jr. and P. B. Heltne, eds., *Neotropical primates: Field studies and conservation*, pp. 101–9. Washington: National Academy of Sciences.

Neyman, P. F. 1978. Aspects of the ecology and social organization of free-ranging cotton-top tamarins (Saguinus oedipus) and the conservation status of the species. In D. G. Kleiman, ed., *The biology and conservation of the callitrichidae*, pp. 39–71. Washington: Smithsonian Institution Press.

Nimuendajú, C. 1948. The Guajá. In J. H. Steward, ed., *The tropical forest tribes*. Vol. 3 of *Handbook of South American Indians, Bureau of American Ethnology Bulletin* 143, pp. 135–36. Washington: Smithsonian Institution Press.

Nishida, T. 1993. Chimpanzees are always new to me. In P. Cavalieri and P. Singer, eds., *The great ape project: Equality beyond humanity*, pp. 24–26. New York: St. Martin's Press.

Nobre de Madeiro, R. A. 1988. Relatório de viagem-Guajá. Ministério da Justiça, Fundação Nacional do Índio.

——. 1990. Relatório de viagem-Guajá. Ministério da Justiça, Fundação Nacional do Índio.

Norconk, M. A., C. Wertis, and W. G. Kinzey. 1997. Seed predation by monkeys and macaws in eastern Venezuela: Preliminary findings. *Primates* 38:177–84.

Noronha, J. M. 1856. Roteiro da viagem da cidade do Pará até ás ultimas colonias dos dominios Portuguezes em os Rios Amazonas, e Negro. In *Collecção de noticias para a história e geografia as naçoes ultramarinas que vivem nos dominos Portuguezes, ou Lhes São Visinhas*, pp. 1–102. Lisbon: Academia Real das Sciencias

North American Liberation Front (NALF). 2002. 2001 year-end direct action report. Courtenay, B.C., Canada: North American Liberation Front Press Office.

Nowak, R. 1994. Xenotransplants set to resume. *Science* 266:1148–51.

Oates, J. F. 1977. The guereza and man: How man has affected the distribution and abundance of *Colobus guereza* and other black colobus monkeys. In His Serene Highness Prince Ranier III of Monaco and G. H. Bourne, eds., *Primate conservation*, pp. 419–67. New York: Academic Press.

O'Brien, S. J., W. G. Nash, D. E. Wildt, M. E. Bush, and R. E. Benveniste. 1985. A molecular solution to the riddle of the giant panda's phylogeny. *Nature* 317:140–44.

Ohnuki-Tierney, E. 1984. Native anthropologists. *American Ethologist* 11:584–86.

——. 1987. *The Monkey as mirror: Symbolic transformations in Japanese history and ritual*. Princeton, N.J.: Princeton University Press.

——. 1990. Monkey as metaphor? Transformations of a polytropic symbol in Japanese culture. *Man* 25:89–107.

Ogden, T. E. 1994. Ophthalmologic research in the owl monkey. In J. F. Baer, R. E. Weller, and I. Kakoma, eds., *Aotus: The owl monkey*, pp. 263–86. San Diego: Academic Press.

Oppenheimer, J. R. 1977. *Presbytis entellus*: The Hanuman languar. In His Serene Highness Prince Ranier III of Monaco and G. H. Bourne, eds., *Primate conservation*, pp. 469–512. New York: Academic Press.

Oppenheimer, J. R. and E. C. Oppenheimer. 1973. Preliminary observations of *Cebus nigrivattatus* (Primates: Cebidae) on the Venezuelan llanos. *Folia Primatologia* 19:409–36.

Ortner, S. B. 1974. Is male to female as nature is to culture? In M. Z. Rosaldo and L. Lamphere, eds., *Women, culture, and society*, pp. 67–88. Stanford: Stanford University Press.

Ottoni, E. B. and M. Mannu. 2001. Semi free-ranging tufted capuchins (*Cebus apella*) spontaneously use tools to crack open nuts. *International Journal of Primatology* 22:347–58.

Panger, M. 1998. Object use in free-ranging white-faced capuchins (*Cebus capucinus*) in Costa Rica. *American Journal of Physical Anthropology* 106:311–21.

——. 1999. Capuchin object manipulation. In P. Dolhinow and A. Fuentes, eds., *The nonhuman primates*, pp. 115–29. Mountain View, Calif.: Mayfield.

Parise, F. 1987. Plano de ação para reestruturação da frente de atração Guajá e programa Awa. Fevereiro. Ministério da Justiça, Fundação Nacional do Índio.

——. 1988. Relatório histórico dos grupos Guajá contatados e sua situação atual. Ministério da Justiça, Fundação Nacional do Índio.

Parise, V. 1973a. Exposição da motivos. Ministério da Justiça, Fundação Nacional do Índio.

——. 1973b. Relatorio da viagem ao Alto Rio Caru e Igarape da Fome para verificar a presença de índios Guajá. Ministério da Justiça, Fundação Nacional do Índio.

Paterson, J. D. 1992. *Primate behavior*. Prospect Height, Ill.: Waveland Press.

Peres, C. 1999. Effects of subsistence hunting and forest types on the structure of Amazonian primate communities. In C. H. Janson and K. E. Reed, eds., *Primate communities*, pp. 268–283. Cambridge: Cambridge University Press.

Peters, C. J., E. D. Johnson, P. B. Jahrling, T. G. Ksiazek, P. E. Rollin, J. White, W. Hall, R. Trotter, and N. Jaax. 1993. Filoviruses. In S. S. Morse, ed., *Emerging viruses*, pp. 159–75. New York: Oxford University Press.

Pickering, M. 1999. Consuming doubts: What some people ate? or what some people swallowed? In L. R. Goldman, ed., *The anthropology of cannibalism*, pp. 51–74. Westport, Conn.: Bergin and Garvey.

Porter, C. A., J. Czelusniak, H. Schneider, M. P. Schneider, I. Sampais, and M. Goodman. 1999. Sequences from the 5' flanking region of the ϵ-globin gene support the relationship of *Callicebus* with the Pithiciins. *American Journal of Primatology* 48:69–75.

Porter, C. A., S. L. Page, J. Czelusniak, H. Schneider, M. P. Schneider, and I. Sampaio. 1997. Phylogeny and evolution of selected primates as determined by sequences of the ϵ-globin locus and 5' flanking regions. *International Journal of Primatology* 18:261–295.

Queiroz, H. L. 1992. A new species of capuchin monkey, genus *Cebus* Erxleben, 1777 (Cebidae: Primates) from eastern Brazilian Amazonia. *Goeldiana (Zoologica)* 15:1–13.

Queiroz, H. L. and R. Kipnis. 1991. Os índios Guajá e os primatas da Amazonia Maranhense. Paper presented at the Congresso Brasileiro de Zoologia. Salvador, Bahía.

Ramirez, M. F., C. H. Freese, and C. J. Revilla 1978. Feeding ecology of the pygmy marmoset, *Cebuella pygmaea*, in northeastern Peru. In D. G. Kleiman, ed., *The biology and conservation of the callitrichidae*, pp. 91–104. Washington: Smithsonian Institution Press.

Rappaport, R. A. 1979. *Ecology, meaning, and religion.* Berkeley: North Atlantic.

Ratcliffe, B. C. 1988. The significance of scarab beetles in the ethnoentomology of non-industrial, indigenous peoples. In D. A. Posey and W. L. Overal, eds., *Ethnobiology: Implications and applications. Proceedings of the First International Congress of Ethnobiology (Belém, 1988).* Vol. 1, pp. 160–85. Belém: Museu Paraense Emílio Goeldi.

Ramos-Elorduy, J. 1988. Edible insects: barbarism or solution to the hunger problem? In D. A. Posey and W. L. Overal, eds., *Ethnobiology: Implications and applications. Proceedings of the First International Congress of Ethnobiology (Belém, 1988).* Vol. 1, pp. 150–57. Belém: Museu Paraense Emílio Goeldi.

Redford, K. H. 1991. The ecologically noble savage. *Cultural Survival Quarterly* 15:46–48.

Renquist, D. M. and R. A. Whitney Jr. 1987. Zoonoses acquired from pet primates. *Veterinary Clinics of North America: Small Animal Practice* 17:219–40.

Richard, A. F., S. J. Goldstein, and R. E. Dewar. 1989. Weed macaques: The evolutionary implications of macaque feeding ecology. *International Journal of Primatology* 10:569–94.

Richards, P. 1993. Natural symbols and natural history, chimpanzees, elephants and experiments in Mende thought. In K. Milton, ed., *Environmentalism: The view from anthropology*, pp. 144–59. New York: Routledge.

Rival, L. 1993. The growth of family trees: Understanding Huaorani perceptions of the forest. *Man* 28:635–52.

——. 1996. Blowpipes and spears: The social significance of Huaorani technological choices. In P. Descola and G. Pálsson, eds., *Nature and society: Anthropological perspectives*, pp. 145–64. New York: Routledge.

——. 1998. Domestication as a historical and symbolic process: Wild gardens and cultivated forests in the Ecuadorian Amazon. In W. Balée, ed., *Advances in historical ecology*, pp. 232–50. New York: Columbia University Press.

Rivière, P. 1969. *Marriage among the Trio: A principle of social organisation.* Oxford: Clarendon.

——. 1984. *Individual and society in Guiana: A comparative study of amerindian social organization.* Cambridge: Cambridge University Press.

Rivero, M. and O. Arredondo. 1991. *Paralouatta varonai*, a new Quaternary platyrrhine from Cuba. *Journal of Human Evolution* 21:1–11.

Rodrigues, A. D. 1984–85. Relações internas na família lingüistica Tupi-Guarani. *Revista de Antropologia* 27–28: 33–53.

Roosevelt, A. C., R. A. Housley, M. Imazio Da Silveira, S. Maranca, and R. Johnson. 1991. Eighth millennium pottery from a prehistoric shell midden in the Brazilian Amazon. *Science* 254:1621–24.

Roosevelt, A. C., M. Lima da Costa, C. Lopes Machado, M. Michab, N. Nercier, H. Valladas, J. Feathers, W. Barnett, M. Imazio da Silveira, A. Henderson, J. Silva, B. Chernoff, D. S. Reese, J. A. Holman, N. Toth, and K. Schick. 1996. Paleoindian cave dwellers in the Amazon: The peopling of the Americas. *Science* 272:373–84.

Rosaldo, M. Z. 1974. Women, culture, and society: A theoretical overview. In M. Z. Rosaldo and L. Lamphere, eds., *Women, culture, and society*, pp. 17–48. Stanford: Stanford University Press.

Rosch, E. 1978. Principles of categorization. In E. Rosch and B. Lloyd, eds., *Cognition and categorization*, pp. 27–48. Hillsdale, N.J.: Lawrence Erlbaum.

Rose, A. L. 2002. Conservation must pursue human-nature biosynergy in the era of social chaos and bushmeat commerce. In A. Fuentes and L. Wolfe, eds., *Primates face to face: The conservation implications of human-nonhuman primate interconnections*, pp. 208–39. Cambridge: Cambridge University Press.

Rosenberger, A. L., W. C. Hartwig, and R. G. Wolff. 1991. *Szalatavus attricuspis*, an early platyrrhine primate. *Folia Primatologia* 56: 225–33.

Rosenberger, A. L., T. Setoguchi, and W. C. Hartwig. 1991. *Laventiana annectens*, new genus and species: Fossil evidence for the origins of Callitrichine New World monkeys. *Proceedings of the National Academy of Sciences of the United States of America* 88:2137–40.

Rosenberger, A. L., T. Setoguchi, and N. Shigehara. 1990. The fossil record of callitrichine primates. *Journal of Human Evolution* 19: 209–36.

Rumsey, A. 1999. White man as cannibal in the New Guinea highlands. In L. R. Goldman, ed., *The anthropology of cannibalism*, pp. 51–74. Westport, Conn.: Bergin and Garvey.

Rylands, A. B., A. F. Coimbra-Filho, and R. A. Mittermeier. 1993. Systematics, geographic distribution, and some notes on the conservation status of the Callitrichidae. In A. B. Rylands, ed., *Marmosets and tamarins: systematics, behaviour, and ecology*, 11–77. Oxford: Oxford University Press.

Rylands, A. B., R. A. Mittermeier, and E. R. Luna. 1997. Conservation of Neotropical primates: threatened species and an analysis of primate diversity by country and region. *Folia Primatologia* 68:134–160.

Sabater Pí, J. and C. Groves. 1972. The importance of higher primates in the diet of the Fang of Río Muni. *Man* 7: 239–43.

Saegusa, A. 1998. Storm in Japan over sale of zoo monkeys for research. *Nature* 393:404.

——. 2000. Congo war increases threat to bonobo research. *Nature* 405:262.

Sahlins, M. D. 1961. The segmentary lineage: An organization of predatory expansion. *American Anthropologist* 63:322–45.

Santos, J. A. 1990. Relatório de Atividades Correspondente a 08 a 09 Feveriro/1990, PINC AWÁ. Ministério da Justiça, Fundação Nacional do Índio.

Sarich, V. M. and J. E. Cronin. 1980. South American mammal systematics, evolutionary clocks, and continental drift. In R. L. Ciochon and A. B. Chiarelli, eds., *Evolutionary biology of the New World monkeys and continental drift*, pp. 339–421. New York: Plenum Press.

Sarich, V. M., C. W. Schmid, and J. Marks. 1989. DNA hybridization as a guide to phylogeny: A critical appraisal. *Cladistics* 5:3–32.

Schapera, I. 1951 [1930]. *The Khoisan peoples of South Africa: Bushmen and Hottentots.* London: Routledge and Kegan Paul.

Scheper-Hughes, N. 1992. *Death without weeping: The violence of everyday life in Brazil.* Berkeley: University of California Press.

Schlichte, H. 1977. A preliminary report on the habitat utilization of a group of howler monkeys (*Alouatta villosa pigra*) in the National Park of Tikal, Guatemela. In T. H. Clutton-Brock, ed., *Primate ecology: Studies of feeding and ranging behaviour in lemurs, monkeys and apes*, pp. 551–59. London: Academic Press.

Schneider, H., M. I. Sampaio, M. P. Schneider, J. M. Ayres, C. M. Barroso, A. R. Hamel, B. T. Silva, and F. M. Salzano. 1991. Coat color and biochemical variation in Amazonian wild populations of *Alouatta belzebul. American Journal of Physical Anthropology* 85:85–93.

Schneider, H. and A. L. Rosenberger. 1996. Molecules, morphology, and platyrrhine systematics. In M. A. Norconk, A. L. Rosenberger, and P. A. Garber, eds., *Adaptive radiations of Neotropical primates*, pp. 3–19. New York and London: Plenum Press.

Seeger, A. 1981. *Nature and society in central Brazil: the Suya Indians of Mato Grosso.* Cambridge: Harvard University Press.

Seymour-Smith, C. 1991. Women have no affines and men no kin: The politics of the Jivaroan gender relations. *Man* 26: 629–49.

Shapiro, J. 1987. From Tupã to the land without evil: The christianization of Tupi-Guarani cosmology. *American Ethnologist* 14:126–39.

Shepard, G. H., Jr. 1997. Monkey hunting with the Machiguenga: Medicine, magic, ecology, and mythology. Paper presented at the American Anthropological Association Invited Session: What Are We Doing Watching Monkeys? Anthropological Perspectives and the Role of Nonhuman Primate Research. November, 1997.

——. 2002. Primates in Matsigenka subsistence and world view. In A. Fuentes and L. Wolfe, eds., *Primates face to face: The conservation implications of human-nonhuman primate interconnections*, pp. 101–36. Cambridge: Cambridge University Press.

Shipman, P., W. Bosler, and K. L. Davis. 1981. Butchering of giant geladas at an Acheulian site. *Current Anthropology* 22:257–68.

Sibal, L. R. and K. J. Samson. 2001. Nonhuman primates: A critical role in current disease research. *Institute for Laboratory Animal Research* 42:74–84.

Sibley, C. G. and J. E. Ahlquist. 1984. The phylogeny of the hominoid primates, as indicated by DNA-DNA hybridization. *Journal of Molecular Evolution* 20:2–15.

——. 1987. DNA hybridization evidence of hominoid phylogeny: Evidence from an expanded data set. *Journal of Molecular Evolution* 26:99–121.

——. 1993. Reviewing misconduct? *American Scientist* 81:407.

Sibley, C. G., J. A. Comstock, and J. E. Ahlquist. 1990. DNA hybridization evidence of hominoid phylogeny: A reanalysis of the data. *Journal of Molecular Evolution* 30:202–36.

Sicotte, P. and P. Uwengeli. 2002. Reflections on the concept of nature and gorillas in Rwanda: Implications for conservation. In A. Fuentes and L. D. Wolfe, eds., *Primates face to face: Conservation implications of human-nonhuman primate interconnections*, pp. 163–81. Cambridge: Cambridge University Press.

Silverwood-Cope, P. 1972. A contribution to the ethnography of the Columbian Maku. Ph.D. Diss., Selwyn College, University of Cambridge.

Simons, E. L. 1997. Preliminary description of the cranium of *Proteopithecus sylviae*, an Egyptian Late Eocene Anthropoidean primate. *Proceedings of the National Academy of Sciences of the United States of America* 94:14,970–75.

Singer, P. 1993. *Practical ethics*. Cambridge University Press.

——. 2001. Heavy petting. Review of M. Dekker, *Dearest pet: On bestiality*. *Nerve* (http://www.nerve.com/opinions/singer/heavypetting), March/April.

Smaglik, P. 2000. NIH takes charge of chimps infected in experiments. *Nature* 405:262.

Small, M. F. 1994. Macaque see, macaque do. *Natural History* 103:8–11.

Smith, C. C. 1977. Feeding behavior and social organization in howling monkeys. In T. H. Clutton-Brock, ed., *Primate ecology: Studies of feeding and ranging behaviour in lemurs, monkeys and apes*, pp. 535–49. London: Academic Press.

Smith, N. 1981. *Man, fishes, and the Amazon*. New York: Columbia University Press.

Smole, W. J. 1976. *The Yanoama Indians: A cultural geography*. Austin: University of Texas Press.

Snowdon, C. T. and P. Soini. 1988. The tamarins, genus *Saguinus*. In R. A. Mittermeier, A. F. Coimbra-Filho, and G. A Fonseca, eds., *Ecology and behavior of Neotropical primates*. Vol. 2, pp. 223–98. Washington: World Wildlife Fund.

Sobral, M. L. 1986. *As missões religiosas e o barroco no Pará*. Belém: Universidade Federal do Pará.

Soini, P. 1972. The capture and commerce of live monkeys in the Amazonian region of Peru. *International Zoo Yearbook* 12: 26–36.

Southwick, C. H. and F. F. Siddiqi. 1977. Population dynamics of Rhesus monkeys in Northern India. In His Serene Highness Prince Ranier III of Monaco and G. H., eds., *Primate conservation*, pp. 339–62. New York: Academic Press.

Speth, J. D. and K. A. Spielmann. 1983. Energy source, protein metabolism, and hunter-gatherer subsistence strategies. *Journal of Anthropological Archaeology* 2:1–31.

Sponsel, L. E. 1997. The human niche in Amazonia: Explorations in ethnoprimatology. In W. G. Kinzey, ed., *New World primates: Ecology, evolution, and behavior*, pp. 143–65. Chicago: Aldine.

Sponsel, L. E., P. Natadecha-Sponsel, and N. Ruttanadakul. In press. Coconut-picking macaques in southern Thailand: Ecological, economic, and cultural aspects. In J. Knight, ed., *Wildlife in Asia: Cultural perspectives*. Richmond, Va.: Curzon Press.

Sponsel, L. E., N. Ruttanadakul, and P. Natadecha-Sponsel. 2002. Monkey business? The conservation implications macaque ethnoprimatology in southern Thailand. In A. Fuentes and L. Wolfe, eds., *Primates face to face: Conservation implications of human-nonhuman primate interconnections*, pp. 288–309. Cambridge: Cambridge University Press.

Sprague, D. S. 2002. Monkeys in the backyard: Encroaching wildlife and rural communities in Japan. In A. Fuentes and L. D. Wolfe, eds., *Primates face to face: Conservation implications of human-nonhuman primate interconnections*, pp. 254–75. Cambridge: Cambridge University Press.

Staden, H. 1928 [1557]. *Hans Staden: The true history of his captivity*. Ed. and trans. Malcolm Letts. London: Routledge.

Stanford, C. B. 1995. The influence of chimpanzee predation on group size and anti-predator behaviour in red colobus monkeys. *Animal Behaviour* 49:577–87.

——. 1996. The hunting ecology of wild chimpanzees: Implications for the evolutionary ecology of Pliocene hominids. *American Anthropologist* 98: 96–113.

——. 1999. *The hunting apes: Meat-eating and the origins of human behavior*. Princeton, N.J: Princeton University Press.

Starlz, T. E., J. Fung, A. Tzakis, S. Todo, A. J. Demetris, I. R. Marino, H. Doyle, A. Zeevi, V. Warty, M. Michaels, S. Kusne, W. A. Rudert, M. Trucco. 1993. Baboon-to-human liver transplantation. *The Lancet* 341:65–71.

Stearman, A. 1989. *Yuquí: Forest nomads in a changing world*. Fort Worth, Tex.: Holt, Rinehart, and Winston.

Stevens, W. K. 1997. Logging sets off an apparent chimpanzee war. New York Times. May 13, 1997.

Stewart, M., V. Pendergast, S. Rumfelt, S. Pierberg, L. Greenspan, K. Glander, and M. Clarke. 1998. Parasites of wild howlers. *International Journal of Primatology* 19: 493–512.

Strum, S. C. 1984. A role for long-term primate field research in source countries. In J. G. Else and P. C. Lee, eds., *Primate ecology and conservation*, pp. 215–20. Cambridge: Cambridge University Press.

Sullivan, N. J., A. Sanchez, P. E. Rollin, Z. Yang, and G. J. Nabel. Development of a preventive vaccine for Ebola virus infection in primates. *Nature* 408: 605–9.

Sutton, M. Q. 1988. Insect resources and Plio-Pleistocene hominid evolution. In D. A. Posey and W. L. Overal, eds., *Ethnobiology: Implications and applications. Proceedings of the first international congress of ethnobiology (Belém, 1988)*. Vol. 1, pp. 195–207. Belém: Museu Paraense Emílio Goeldi.

Sweet, D. G. 1975. A rich realm of nature destroyed: The middle Amazon valley, 1640–1750. Ph.D. diss., University of Wisconsin. Ann Arbor: Microfilms International.

Symmes, D. and P. Goedeking. 1988. Nocturnal vocalizations by squirrel monkeys (*Saimiri sciureus*). *Folia Primatologia* 51:143–48.

Takai, M. 1994. New specimens of *Neosaimiri fieldsi* from La Venta, Colombia: A middle Miocene ancestor of the living squirrel monkey. *Journal of Human Evolution* 27:329–60.

Takai, M., F. Anaya, N. Shigehara, and T. Setoguchi. 2000. New fossil materials of the earliest *Branisella boliviana* and the problem of platyrrhine origins. *American Journal of Physical Anthropology* 111:263–81.

Taylor, A. C. 1996. The soul's body and its states: An Amazonian perspective on the nature of being human. *Journal of the Royal Anthropological Institute* 2:201–15.

——. 1998. Jivaro kinship: "Simple" and "Complex" Formulas: A Dravidian transformation group. In M. Godelier, T. R. Trautmann, and F. E. Tjon Sie Fat, eds., *Transformations of kinship*, pp. 187–213. Washington: Smithsonian Institution Press.

——. 2001. Wives, pets, and affines: Marriage among the Jivaro. In N. L. Whitehead and L. M. Rival, eds., *Beyond the visible and material: The Amerindianization of society in the work of Peter Rivière*, pp. 45–56. Oxford: Oxford University Press.

Teleki, G. 1993. They are us. In P. Cavalieri and P. Singer, eds., *The great ape project: Equality beyond humanity*, pp. 296–302. New York: St. Martin's Press.

Terborgh, J. 1983. *Five New World primates: A study in comparative ecology*. Princeton, N.J.: Princeton University Press.

Thorington, R. W., Jr., N. A. Muckenhirn, and G. G. Montgomery. 1976. Movements of a wild night monkey (*Aotus trivirgatus*). In R. W. Thorington Jr. and P. G. Heltne, eds., *Neotropical primates*, pp. 32–35. Washington: National Academy of Sciences.

Tomaz Filho, F. P. 1991a. Relatório de atividades PINC Awá, mes Abril 1991. Ministério da Justiça, Fundação Nacional do Índio.

——. 1991b. Relatório de atividades PINC Awá, mes Maio 1991. Ministério da Justiça, Fundação Nacional do Índio.

——. 1991c. Relatório de atividades PINC Awá, mes Junho 1991. Ministério da Justiça, Fundação Nacional do Índio.

Tonkinson, R. 1991. *The Mardu Aborigines: Living the dream in Australia's desert*. 2d edition. Fort Worth, Tex.: Holt, Rinehart, and Winston.

Toral, A. A. 1991. A saga de Karipiru. In *Povos Indígenas no Brasil 1987/88/89/90*, pp. 362–64. São Paulo: Centro Ecumênico de Documentação e Informação.

Torres de Assumpção, C. 1981. *Cebus apella* and *Brachyteles arachnoides* (Cebidae) as potential pollinators of *Mabea fistulifera* (Euphorbiaceae). *Journal of Mammalogy* 62:386–88.

Townsend, P. K. 1990. On the possibility/impossibility of tropical forest hunting and gathering. *American Anthropologist* 92:745–47.

Trautmann, T. R. 1981. *Dravidian Kinship*. Cambridge: Cambridge University Press.

Turner, T. S. 1979. The Gê and Bororo societies as dialectical systems: A general model. In D. Maybury-Lewis, ed., *Dialectical societies: The Gê and Bororo of central Brazil*, pp. 147–78. Cambridge: Harvard University Press.

Tylor, E. B. [1871] 1958. *Primitive culture*. New York: Harper Torchbooks.

Urbani, B. 1999. Nuevo mundo, nuevos monos: Sobre primates Neotropicales en los siglos XV y SVI. *Neotropical Primates* 7:121–25.

Urbani, B. and E. Gil. In press. *Consideraciones sobre restos de primates de un yacimento arqueológico del oriente de Venezuela (América del Sur): Cueva del Guácharo, estado Monagas*.

van Roosmalen, M. G., R. A. Mittermeier, and J. Fleagle. 1988. Diet of the northern bearded saki (*Chiropotes satanas chiropotes*): A Neotropical seed predator. *American Journal of Primatology* 14:11–35.

van Roosmalen, M. G., R. A. Mittermeier, and K. Milton. 1981. The bearded Sakis, genus *Chiropotes*. In A. F. Coimbra-Filho and R. A. Mittermeier, eds., *Ecology and behavior of Neotropical primates*. Vol. 1, pp. 419–41. Rio de Janeiro: Academia Brasileira de Ciências.

Vickers, W. T. 1988. Game depletion hypothesis of Amazonian adaptation: Data from a native community. *Science* 239:1521–22.

Vilaça, A. 1992. *Comendo como gente: Formas do canibalismo Wari'*. Rio de Janeiro: Universidade Federal do Rio Janeiro.

Visalberghi, E. and D. M. Fragaszy. 1990. Do monkeys ape? In S. Parker and K. Gibson, eds. *"Language" and intelligence in monkeys and apes*, pp. 247–73. Cambridge: Cambridge University Press.

Visalberghi, E. and W. C. McGrew. 1997. *Cebus* meets *Pan. International Journal of Primatology* 18:677–81.

Visalberghi, E. and L. Trinca. 1989. Tool use in capuchin monkeys, or distinguishing between performing and understanding. *Primates* 30:511–21.

Viveiros de Castro, E. 1992. *From the enemy's point of view: Humanity and divinity in Amazonian society*. Trans. Catherine V. Howard. Chicago: The University of Chicago Press.

——. 1998a. Cosmological deixis and Amazonian perspectivism. *Journal of the Royal Anthropological Institute* 4:469–88.

——. 1998b. Dravidian and related kinship systems. In M. Godelier, T. R. Trautmann, and F. E. Tjon Sie Fat, eds., *Transformations of kinship*, pp. 332–85. Washington: Smithsonian Institution Press.

——. 1999a. Comments to Nurit Bird David's " 'Animism' revisited." *Current Anthropology* 40:S79.

——. 1999b. The transformation of objects into subjects in Amerindian ontogenies. Paper presented at the American Anthropological Association invited session: Reanimating Religion: A Debate on the New Animism. Chicago. November.

——. 2001. Gut feelings about Amazonia: Potential affinity and the construction of sociality. In N. L. Whitehead and L. M. Rival, eds., *Beyond the visible and material: The Amerindianization of society in the work of Peter Rivière*, pp. 19–43. Oxford: Oxford University Press.

Vogt, J. L. 1978. The social behavior of a marmoset (*Saguinus fuscicollis*) group: III, spatial analysis of social structure. *Folia Primatologia* 29:250–67.

Vogt, J. L., H. Carlson, and E. Menzel. 1978. Social behavior of a marmoset (*Saguinus fuscicollis*) group: I, parental care and infant development. *Primates* 19:715–26.

Von Graeve, B. 1989. *The Pacaa Nova: Clash of cultures on the Brazilian frontier*. Peterborough, Ont.: Broadview Press.

Wagley, C. [1977] 1983. *Welcome of tears: The Tapirapé Indians of central Brazil*. Prospect Heights, Ill.: Waveland Press.

Wallis, J. and D. R. Lee. 1999. Primate conservation: The prevention of disease transmission. *International Journal of Primatology* 20: 803–26.

Weller, R. E. 1994. Infectious and non-infectious diseases of owl monkeys. In J. F. Baer, R. E. Weller, and I. Kakoma, eds., *Aotus: The owl monkey*, pp. 177–215. San Diego: Academic Press.

Werner, D. 1984. *Amazon journey: An anthropologist's year among Brazil's Mekranoti Indians*. New York: Simon and Schuster.

Westergaard, G. C. and S. J. Suomi. 1997a. Capuchin monkey (*Cebus apella*) grips for the use of stone tools. *American Journal of Physical Anthropology* 103:131–35.

——. 1997b. Transfer of tools and food between groups of tufted capuchins (*Cebus apella*). *American Journal of Primatology* 43:33–41.

Wheatley, B. P. 1999. *The sacred mokeys of Bali*. Prospect Heights, Ill.: Waveland Press.

Wheatley, B. P., R. Stephenson, H. Kurashina, and K. G. Marsh-Kautz. 1997. A preliminary study of the effects of human hunting on *Macaca fascicularis* of the Ngeaur Island, Republic of Belau. Paper presented at the American Anthropological Association invited session: What Are We Doing Watching Monkeys? Anthropological Perspectives and the Role of Nonhuman Primate Research. November, 1997.

——. 2002. A cultural primatological study of *Macaca fascicularis* on Ngeaur Island, Republic of Palau. In A. Fuentes and L. Wolfe, eds., *Primates face to face: The conservation implications of human-nonhuman primate interconnections*, pp. 240–53. Cambridge: Cambridge University Press.

Whitehead, N. L. 2001. Kanaimà: Shamanism and ritual death in the Pakaraima Mountains, Guyana. In N. L. Whitehead and L. M. Rival, eds., *Beyond the visible and material: The Amerindianization of society in the work of Peter Rivière*, pp. 235–45. Oxford: Oxford University Press.

Whiten, A., J. Goodall, W. C. McGrew, T. Nishida, V. Reynolds, Y. Sugiyama, C. E. Tutin, R. W. Wrangham, and C. Boesch. 1999. Cultures in chimpanzees. *Nature*: 682–85.

Wiebers, D. O., A. Elzanowski, P. W. Gikas, J. Leaning, and R. D. White. 1997. Monkey business in space. *Nature* 389:537.

Wieczkowski, J. and D. N. Mbora. 2002. Increasing threats to the conservation of endemic endangered primates and forests of the Lower Tana River, Kenya. *African Primates* 4:32–40.

Winter, P. 1969. The variability of peep and twit calls in captive squirrel monkeys (*Saimiri sciureus*). *Folia Primatologia* 10:205–15.

Wolfe, L. D. 1991. Macaques, pilgrims, and tourists re-visited. *National Geographic Research and Exploration* 7:241.

——. 1992. Feeding habits of the Rhesus monkeys (*Macaca mulatta*) of Jaipur and Galta, India. *Human Evolution* 7:43–54.

——. 1997. The State of Florida vs. the monkeys of Silver Springs. Paper presented at the American Anthropological Association Invited Session: What Are We Doing Watching Monkeys? Anthropological Perspectives and the Role of Nonhuman Primate Research. November, 1997.

———. 2002. Rhesus macaques: A comparative study of two sites, Jaipur, India, and Silver Springs, Florida. In A. Fuentes and L. Wolfe, eds., *Primates face to face: The conservation implications of human-nonhuman primate interconnections*, pp. 310–30. Cambridge: Cambridge University Press.

Wolfe, L. D. and J. P. Gray. 1982. Japanese monkeys and popular culture. *Journal of Popular Culture* 16:97–105.

Wolfheim, J. H. 1983. *Primates of the world: Distribution, abundance, and conservation*. Seattle: University of Washington Press.

Woo, J. K. 1966. The skull of Lantian Man. *Current Anthropology* 7:83–86.

Wood, A. E. 1980. The origin of the caviomorph rodents from a source in Middle America: A clue to the area origin of platyrrhine primates. In R. L. Ciochon and A. B. Chiarelli, eds., *Evolutionary biology of the New World monkeys and continental drift*, pp. 79–91. New York: Plenum Press.

Wrangham, R. W. and E. V. Bergmann Riss. 1990. Rates of predation on mammals by Gombe Chimpanzees, 1972–1975. *Primates* 31:157–70.

Wright, P. 1978. Home range, activity pattern, and agonistic encounters of a group of night monkeys (*Aotus trivirgatus*) in Peru. *Folia Primatologia* 29:43–55.

———. 1981. The night monkeys, genus *Aotus*. In A. F. Coimbra-Filho and R. A. Mittermeier. eds., *Ecology and behavior of Neotropical primates*. Vol. 1, pp. 211–40. Rio de Janeiro: Academia Brasileira de Ciências.

———. 1984. Biparental care in *Aotus trivirgatus* and *Callicebus moloch*. In M. F. Small, ed., *Female primates: Studies by women primatologists*, pp. 59–72. New York: Allan R. Liss.

———. 1996. The Neotropical primate adaptation to nocturnality: Feeding in the night (*Aotus nigriceps* and *A. azarae*). In M. A. Norconk, A. L. Rosenberger, and P. A. Garber, eds., *Adaptive radiations of Neotropical primates*, pp. 369–82. New York and London: Plenum Press.

Yost, J. A. and P. M. Kelley. 1983. Shotguns, blowguns, and spears: The analysis of technological efficiency. In R. B. Hames and W. T. Vickers, eds., *Adaptive responses of native Amazonians*, pp. 189–224. New York: Academic Press.

Zeuner, F. 1963. *A History of Domesticated Animals*. New York: Harper and Row.

Zucker, E. L. and M. R. Clarke. 1998. Agonistic and affiliative relationships of adult female howlers (*Alouatta palliata*) in Costa Rica over a 4-year period. *International Journal of Primatology* 19:433–49.

INDEX

Page numbers in **bold** type refer to illustrations.